Progress in STANDARDIZATION of AQUATIC TOXICITY TESTS

Progress in STANDARDIZATION of AQUATIC TOXICITY TESTS

Edited by
Amadeu M.V.M. Soares, University of Coimbra, Portugal
Peter Calow, University of Sheffield, England

SETAC Special Publications Series

Series Editors

Dr. Thomas LaPoint
The Institute of Wildlife and Environmental Toxicology
Clemson University, U.S.A.
Dr. Peter W. Greig-Smith
Fisheries Laboratory
Ministry of Agriculture, Fisheries and Food, United Kingdom

LEWIS PUBLISHERS
Boca Raton Ann Arbor London Tokyo

Library of Congress Cataloging-in-Publication Data

Progress in standardization of aquatic toxicity tests / edited by
Amadeu M. V. M. Soares, Peter Calow.
p. cm.
Includes bibliographical references and index.
ISBN 0-87371-845-3
1. Water quality bioassay--Standards--Congresses. 2. Toxicity testing--Congresses. I. Soares, Amadeu M. V. M. II. Calow, Peter.
QH90.57.B5P76 1993
628.1′61--dc20 92-43053
CIP

Direct all inquiries to CRC Press, Inc., 2000 Corporate Blvd., N.W., Boca Raton, Florida 33431.

PRINTED IN THE UNITED STATES OF AMERICA
1 2 3 4 5 6 7 8 9 0

Printed on acid-free paper

THE SETAC SPECIAL PUBLICATIONS SERIES

The SETAC Special Publications Series was established by the Society of Environmental Toxicology and Chemistry to provide in-depth reviews and critical appraisals on scientific subjects relevant to understanding the impacts of chemicals and technology on the environment. The series consists of single- and multiple-authored/edited books on topics selected by the SETAC Board of Directors and the Council of SETAC-Europe for their importance, timeliness, and their contribution to multidisciplinary approaches to solving environmental problems. The diversity and breadth of subjects covered in the series will reflect the wide range of disciplines encompassed by environmental toxicology, environmental chemistry, and hazard/risk assessment. Despite this diversity, the goals of these volumes are similar; they are to present the reader with authoritative coverage of the literature, as well as paradigms, methodologies, controversies, research needs, and new developments specific to the featured topics. All books in the series are peer reviewed for SETAC by acknowledged experts.

The SETAC Special Publications will be useful to environmental scientists in research, research management, chemical manufacturing, regulation, and education, as well as students considering careers in these areas. The series will provide information for keeping abreast of recent developments in familiar areas and for rapid introduction to principles and approaches in new subject areas.

Thomas W. LaPoint
The Institute of Wildlife and Environmental Toxicology
Clemson University, U.S.A.

Peter W. Greig-Smith
Fisheries Laboratory
Ministry of Agriculture, Fisheries, and Food, U.K.

Series Editors, SETAC Special Publications

PREFACE

Standardization of tests is a major concern for ecotoxicologists. In the past few years, much attention and effort have been devoted to the reduction of variability in ecotoxicological tests. From several interlaboratory exercises (ring- or round-robin tests) and workshops that have been organized at national and international levels, it is apparent that lack of consistency in results may either be the consequence of an inappropriate use of the biological knowledge of test organisms or a lack of standardization within test protocols (often allowing too much flexibility).

There is always a risk that standardization discourages creativity and restrains the development of new test approaches. However, if ecotoxicological tests are to be used as regulatory tools, some standardization is needed. The contributions presented here aim to show how this can be better achieved. They arise out of a SETAC-Europe Meeting held in Sheffield, England, April 1991. All have been subjected to peer review and not all material submitted was accepted. We thank the authors and reviewers for their indulgence and hard work in making this volume come to life. We now hope that the reader will appreciate it as well.

Amadeu M.V.M. Soares graduated in biology from Coimbra University, Portugal, in 1984 and was awarded his Ph.D. for work in aquatic ecotoxicology at Sheffield University, U.K., in 1989. He is currently professor of Toxicology, and Environmental Physiology, at the Department of Zoology, Coimbra University. He is also Coordinator of a research project at the European level, investigating the development of bioassay systems to test poorly water-soluble substances. His research interests include mechanisms of toxicity, methods of toxicity testing and side effects of cyanobacterial blooms, and of mining on freshwater communities. He is one of the founder-members of SETAC-Europe, and was the president of the organizing committee of the first SETAC World Congress. He is also a member of the Portuguese Association for Water Resources, the Portuguese Association of Biologists, and the Scientific Advisory Committee of the European Centre of Validation of Alternative Methods at Ispra, Italy.

Peter Calow is Director of the Institute of Environmental Sciences and Technology, and Professor of Zoology at the University of Sheffield. His research interests have ranged from ecology and evolutionary biology to aquatic ecotoxicology. One of his main concerns at present is in the refinement of prediction test systems and in exploring their relevance to ecological systems.

He has been President of SETAC U.K. and SETAC Europe and is currently Chairman of the U.K. Government's Advisory Committee on Hazardous Substances. He is also the editor of both *Functional Ecology* and *Integrated Environmental Management*. He has published over 150 articles and several books.

Professor Calow obtained his Bachelor's degree, Ph.D., and D.Sc. from the University of Leeds. He spent several years at the University of Glasgow before moving to Sheffield.

CONTRIBUTORS

Luc Brendonck, Ph.D.
Laboratory for Biological Research in Aquatic Pollution
University of Ghent
Ghent, Belgium
Royal Belgian Institute of Natural Sciences
Dept. of Freshwater Biology
Brussels, Belgium

Peter Calow, Ph.D.
Institute for Environmental Science and Technology
University of Sheffield
Sheffield, England

Argelia Castaño, Ph.D.
Department of Animal Health
CIT — INIA
Valdeolmos, Madrid, Spain

Maria Dulce Centeno, Ph.D.
Laboratory for Biological Research in Aquatic Pollution
University of Ghent
Ghent, Belgium

Göran Dave, Ph.D.
Department of Zoophysiology
University of Göteborg
Göteborg, Sweden

R. Dermott, M.Sc.
Great Lakes Laboratory for Fisheries and Aquatic Sciences
Fisheries and Oceans Canada
Canada Centre for Inland Waters
Burlington, Ontario, Canada

Fiona M. Fairhurst
Shell Research Ltd.
Sittingbourne Research Center
Sittingbourne, Kent, England

Andrew E. Girling
Shell Research Ltd.
Sittingbourne Research Center
Sittingbourne, Kent, England

C. R. Janssen, Ph.D.
Laboratory for Biological Research in Aquatic Pollution
University of Ghent
Ghent, Belgium

M. Munawar, Ph.D.
Great Lakes Laboratory for Fisheries and Aquatic Sciences
Fisheries and Oceans Canada
Canada Centre for Inland Waters
Burlington, Ontario, Canada

Guido Persoone, Ph.D.
University of Ghent
Laboratory for Biological Research in Aquatic Pollution
Ghent, Belgium

Amadeu M. V. M. Soares, Ph.D.
Department of Zoology
University of Coimbra
Coimbra, Portugal

J. V. Tarazona, Ph.D.
Department of Animal Health
CIT — INIA
Valdeolmos, Madrid, Spain

Frieda B. Taub, Ph.D.
School of Fisheries HF-15
College of Ocean and Fishery Science
University of Washington
Seattle, Washington

Marleen Van Steertegem, Ph.D.
Laboratory for Biological Research in Aquatic Pollution
University of Ghent
Ghent, Belgium

Graham F. Whale
Shell Research Ltd.
Sittingbourne Research Center
Sittingbourne, Kent, England

CONTENTS

CHAPTER 1

Seeking Standardization in Ecotoxicology

Amadeu M.V.M. Soares and Peter Calow

ABSTRACT

This introductory chapter argues for the primacy of standardization over realism in the development of ecotoxicological tests and, by referring to the other chapters, indicates what is necessary for achieving this. Sources of variation in tests involving individuals and populations derive from both genotype and the environment. As well as these, sources of variation in multispecies systems derive from species composition at the initiation and then throughout the test. Of course, in the end, the results from standard tests have to be related to the responses of real ecological systems. In doing this and also in achieving standardization, much needs to be known about the biology of the test system.

0-87371-845-3/93/$0.00 + $.50

KEY WORDS

Standardization, repeatability, reproducibility, genotype, environment, micro/mesocosm, diversity, realism, relevance.

INTRODUCTION

A tacit presumption of any legal device is that it can be applied uniformly and, if necessary, can be demonstrated. Hence, ecotoxicological tests that are used in legal controls of chemicals must, when applied to the same chemical, give the same result in different laboratories and in the same laboratory at different times. In a later chapter by Dave (Chapter 9) these are referred to respectively as repeatability and reproducibility.

Of course, when measurements are made on the same entity in different places or at different times, they are unlikely to be exactly equivalent because of observer error. Moreover, if the measuring device is biological, the potential sources of variation increase considerably. With systems involving individuals and populations, these can be classified according to the scheme used in classical genetics[1] as those due to genotype, those due to environment, and those due to an interaction between genotype and environment. With systems involving multispecies groupings — communities/ecosystems — as well as these sources of variation, there are those due to species diversity expressed in terms of both the types of organisms present and their relative abundance.

The use of ring tests (round robin tests) indicates that the level of variation attributed to both repeatability and reproducibility is considerable; although, as might be expected, Dave shows that variability involved in reproducibility is usually greater than that involved in repeatability. We currently do not have any formal view about how much variability might be tolerable from the point of view of both legal and scientific credibility, but it is clear that orders of magnitude differences in results involving the application of the same test to the same chemical, as have been reported for the daphnid system (Chapter 4), are not tolerable. Hence, some attention has to be given to reducing the variability by ensuring that tests are run in more strict similar ways, at different times, and in different places. This is what is meant by "standardization" and was the motivation behind the scientific meeting that generated most of the papers on which this book is based (Session in Founding Conference of SETAC-Europe, Sheffield, April 1991). All the papers, it should be said, have been subjected to the normal processes of peer review and not all the papers submitted were ultimately included.

SOURCES OF VARIATION IN TESTS INVOLVING INDIVIDUALS AND POPULATIONS

As already indicated, the major potential sources of variation in tests involving individuals and populations are genetic and environmental. The parthenogenetic reproduction of *Daphnia*, and hence its clonal organization, provide a convenient way of teasing these two components apart (Chapter 4) and do indicate that genotype matters. Some genotypes are more tolerant to toxicants than others, but unfortunately there do not appear to be generally tolerant or sensitive genotypes. This leads to the dilemma of which to use (see below).

Several environmental factors can be controlled with relative ease and precision such as temperature and lighting regimes. Equally, there are many factors that are difficult to control and yet potentially important in affecting the quality of the test organisms and hence their response to toxicants. Chief amongst these are held and the supply of live food. Moreover, their conditions are not just important in the immediate circumstances of the test, but in stock cultures used to generate test animals. For example, female daphnids contain three discernible generations of offspring that can be potentially open to maternal influence: ones in a brood pouch (broodlings), ones in the ovary that are being provisioned and awaiting deposition in the brood pouch upon release of broodlings, and ones in the ovary awaiting provisioning.[2] The maternal condition influences provisioning, and this in turn can potentially influence the quality of the offspring.[3] Hence, maternal effects in a culture might have implications for the ecotoxicological responses of three generations of offspring produced from it.

It is for reasons such as these, and also the considerable labor needed to service cultures, that much effort has been put into the development of cyst-based toxicity tests by Persoone et al. (Chapters 2, 3, and 6). The resting stages in the life cycles of several aquatic invertebrates provide convenient material that can be stored "on the shelf" without the need for feeding and culturing until required. Similarly, the use of non-feeding early life stages (Chapter 9), cell lines (Chapter 8), and other *in vitro* systems can remove these sources of variation. At the same time, potential maternal effects, of the kind described for daphnids, do mean that care has to be taken in collecting eggs, embryos, and cysts for the purpose of running tests.

SOURCES OF VARIATION IN MULTISPECIFIC TESTS

Laboratory multispecies test systems can either be derived from nature or by construction. The former involves sampling from systems that are naturally variable in space and time and so standardization of initial conditions is usually impossible. Moreover, the control of community structure and function under subsequent laboratory conditions is also likely to be difficult.

The alternative — construction of communities from defined components in the laboratory — in principle does not suffer from the problem of not guaranteeing initial conditions, though the subsequent control of components is usually difficult. Taub has taken this approach further than most in aquatic ecology. In Chapter 10, she reports from an inter-laboratory comparison that indicates that at least the trends in both community structure and function are remarkably similar in control and toxicant-stressed constructed microcosms across laboratories.

What cannot be ignored in these systems, though, is the massive amount of labor needed to establish and monitor them.

STANDARDIZATION AND RELEVANCE

The simpler the system the easier it is to standardize; the *in vitro* tests involving enzymes and cell lines (Chapters 5 and 8) are in principle more easily controlled and hence standardized than multispecies micro/mesocosms (Chapter 10). This is the case because simpler systems are, by definition, controlled by fewer variables than complex ones. Similarly, standardization generally involves simplification, e.g., in single species tests use one genotype in a simple, defined medium with a defined, artificial diet.

Yet nature is a complex place and so the question can be asked if by moving towards more standardization we move away from ecological realism.[4] There are two responses to this. First, unless we work with systems that are under control experimentally we cannot know how to interpret the results that derive from them. Second, we must always be asking how the responses that we observe in tests relate to ecological systems — suborganismic responses to population ones,[5] one genotype/species response to those of others (Chapters 4 and 7), and single species to multispecies responses.[6]

As a "bottom line", it seems clear that standardization and experimental control have to take precedence over realism, because without them we have no basis for scientific understanding. What is still not clear is how far we need to push standardization and reduce variability in order to obtain ecotoxicologically useful results. In Chapter 9, Dave indicates how variability can be studied, but now we need to give some attention to how much variability we should generally be targeting.

IMPORTANCE OF BASIC BIOLOGY

In order to be able to standardize and optimize test cultures and test systems, a considerable amount of effort has to be invested in understanding the basic biology of the test organisms or communities. To supply standard, fully-defined media or artificial diets that do not stress test organisms requires a considerable amount of information on their "salt", water, and nutritional requirements (Chapters 2 and 4). To construct stable multispecies systems requires knowledge of the multispecies dynamics and the interactions that lie behind them (Chapter 10). Over and above this, if we are going to be able to generalize from toxicological observations made in the laboratory to field situations, we need to know more about processes of seasonality of toxicity (Chapter 11), toxic mechanisms, and the relationship between the susceptibility of the observational system (be it genotype, species, or communities) and multigenotype/species assemblages in nature.

There is much information in the literature on aspects of these requirements and this needs collecting together, at least for the key test systems as has been started for daphnids (Chapter 4) and other freshwater invertebrates (Chapter 2). Undoubtedly there will be a need for basic research to fill in the gaps.

REFERENCES

1. Falconer, D.G. 1981. *Introduction to Quantitative Genetics.* Longman, Essex, 2nd Edition. 340pp.
2. Bradley, M.C., D.J. Baird and P. Calow. Mechanisms of energy allocation to reproduction in the cladoceran *Daphnia magna* Straus. *Biol J. Linn. Soc.,* 44:325-333.
3. Naylor, C., E.J. Cox, M.C. Bradley and P. Calow. 1992. Effect of defining maternal food rations on susceptibility of *Daphnia magna* Straus neonates to toxic substances. *Aquat. Toxicol.* 24:63–74.
4. Forbes, V.E. and M.H. Depledge. 1992. Predicting population response to pollutants: the significance of sex. *Funct. Ecol.* 6:376–381.
5. Calow, P. and R. Sibly. 1990. A physiological basis of population processes: ecotoxicological implications. *Funct. Ecol.,*4:283-288.
6. Maltby, L. and P. Calow. 1990. The application of bioassays in the resolution of environmental problems; past, present and future. *Hydrobiologia* 188/189, 65-76.

CHAPTER 2

Biological/Ecological Characteristics of Large Freshwater Branchiopods from Endorheic Regions and Consequences for Their Use in Cyst-Based Toxicity Tests

Luc Brendonck and Guido Persoone

ABSTRACT

Most freshwater branchiopods of (semi-) arid regions inhabit lentic waters of a temporary or astatic nature, withstand the dry phase as a resistant egg, and are generally strongly r-selected to maximize the number of offspring in their highly unpredictable habitat. A unique combination of reproductive traits (both qualitative and quantitative) makes these organisms attractive for the development of cyst-based bioassays which bypass all efforts and difficulties inherent in the recruitment or continuous culturing of live stocks. The potential of these

0-87371-845-3/93/$0.00 + $0.50

freshwater branchiopods is examined in the framework of the development of cost-effective toxicity tests for routine toxicity screening of chemicals or effluents.

KEY WORDS

Large freshwater branchiopods, (semi-) arid regions, biological/ecological characteristics, cyst-based toxicity tests.

FIGURE ABBREVIATIONS

populat. = population; red. = reduced; incr. = increased; genet. = genetic; variabil. = variability; supernum. = supernumerary; start mat. = starting material; compet. = competitive; reprod. = reproductive.

INTRODUCTION

The number of species presently used in standardized aquatic toxicity tests remains limited and stands in contrast with the taxonomic richness of most natural systems. This poverty in standard tests is primarily due to the fact that most bioassays imply continuous culturing and/or recruitment of live stocks of the test organisms, in a healthy state and in sufficient numbers.

The concept of using resting (dormant) stages of particular species as "starting" biological material in toxicity testing offers an ideal solution for these problems.[1–3] Although some of the benefits are species-dependent, the general advantage of using resting stages is that they can be stored and hatched at will, thus eliminating the need for culturing stocks with their associated costs.

The brine shrimp *Artemia* is the first organism ever used in toxicity tests starting from resting eggs (so-called cysts). The commercial availability of dried cysts is the main reason why *Artemia* has been used in aquatic bioassays over the past 30 years. A detailed review on the use of *Artemia* in research and applied toxicology is presented in Persoone and Wells[1] and in Van Steertegem and Persoone.[4] Many other organisms also produce resting eggs: turbellarians, gastrotrichs, monogonont rotifers, freshwater branchiopod crustaceans (including the cladocerans), calanoid and ostracod crustaceans, and several groups of insects. Resting stages have also been reported from annual fishes occurring in temporary waters in Africa.[5,6]

In spite of the considerable advantages associated with the use of cyst-derived test-species, only few attempts have thus far been made to develop additional "resting stage"-based toxicity tests as was done for protozoans (cysts),[7,8] rotifers (resting eggs),[2,3,9–12] and daphnids (ephippia + resting egg[s]).[13–16] The majority of species belonging to these groups have a predominantly asexual life cycle with parthenogenetic production of active offspring. Only in response to particular environmental conditions (food, photoperiod, temperature, oxygen, etc.), which precede the harsh extreme, will sexual production of resting stages be induced.[17]

In the majority of non-cladoceran (large) freshwater branchiopods (Anostraca [fairy shrimps], Notostraca [tadpole shrimps], Laevicaudata and Spinicaudata [clam shrimps]), however, no distinct life-cycle phases occur and resting eggs are continuously produced. This group thus presents an important source for the development of cost-effective cyst-based bioassays. Large freshwater branchiopods have long been considered as biological curiosities, since these ancestral crustaceans are largely limited to lentic waters of a temporary or astatic nature. Only recently has there been an upsurge of interest in the use of these organisms for research and various types of applications.[18–23, 26] In particular, the representatives of arid/semi-arid regions are characterized by a unique set of reproductive traits which makes them suitable for applications in aquatic toxicology and aquaculture.

This chapter discusses some general biological and ecological characteristics of large freshwater branchiopods, describing some important features of their ephemeral habitat which, through the reproductive pattern,[24] has fashioned the ecology of the species. Emphasis will be placed on some biotic and life cycle adaptations that maximize the benefits of life in temporary waters in arid and semi-arid, so-called endorheic regions, with special reference to the subtropical fairy shrimp *Streptocephalus proboscideus* (Frauenfeld) and to those adaptations that are relevant for applications in aquatic toxicology.

The systematics, morphology and distribution, the feeding biology, the reproductive biology, some hatching characteristics of the resting eggs, and controlled production of resting eggs of *Streptocephalus proboscideus* have been dealt with in previous papers.[21,25–30] The standardization of an acute cyst-based test for this species will be discussed elsewhere.[23]

SEVERAL ASPECTS OF THE EPHEMERAL HABITAT AND THEIR CONSEQUENCES

Although some members of the fairy, clam, and tadpole shrimps can occupy littoral or even pelagic regions of large permanent lakes,[31–33] non-cladoceran freshwater branchiopods typically inhabit temporary, aquatic environments that (ir)regularly dry out or show striking changes in the water level providing so-called "enemy-free space". Such habitats predominantly occur in (semi-) arid regions and temperate climates.[34] Large freshwater branchiopods are, nevertheless, most abundant in arid and semi-arid regions.

To date, several classifications have been used to discriminate between different types of ephemeral pools. Wiggins et al.,[17] for example, demarcated different sorts of temporary pools by length of wet and dry phases, marginal vegetation, and basin vegetation. Decksbach[35] distinguished between seasonally astatic waters that dry up annually and perennially astatic waters that display a marked annual fluctuation in level without drying out. Williams[36] considered six major types of temporary water bodies on a combination of the degree of their predictability, the salinity level, and the extent to which these waters occupy discrete basins or are associated with seasonally or unpredictably flooding rivers. Morgan and Boy[37] constructed a general classification of wetlands and open waters of Northwest Africa using physical, botanical, and faunal data, while Weir[38] grouped Central African rainpools based on the level of pollution and the presence of dung, vegetation, and organic sediment.

The hazardous nature of the temporary habitat, depending on rainfall and disastrous droughts, is the most important factor determining the life cycle of rainpond species from (semi-) arid regions. Some rainpools may be dry for several years,[34,36] while others may last only a few hours.[39] *S. proboscideus*, for example, has been collected from discrete desert freshwater pools near Khartoum (Sudan). Rainfall in this area is unpredictable and pools may fill for only a few weeks or even days.[25,29,40]

In particular, the recurrent droughts (density independent factors) result in limitation on the long-term population growth of large freshwater branchiopods (which includes a free-swimming and a dormant component) by frequent catastrophic mortalities in the active component of the population. Rainfall may even be insufficient for immature specimens to reach the adult stage before the pool dries ("abortive hatching"), resulting in a considerable reduction of the cyst bank. Several papers on field studies report that drying occurred before the organisms attained sexual maturity.[41–45] Density-dependent factors such as predation,[46] accumulation of metabolites, and resource limitation generally only become important in later successional stages in rain-ponds.

Apart from desiccation, additional important ecological stressors in temporary waters in (semi-) arid regions are chemical variation, high temperatures, low oxygen concentration, high light intensities, instability (unpredictability + temporary nature), habitat isolation, and variable water levels.[17,32,34,36,47] It must be noted, however, that stressors generally act in concert and consequently adaptations to one stressor alone are unlikely to ensure survival: adaptation is mostly a multi-dimensional phenomenon.[48]

Plentiful food seems to be a general feature of temporary rainpools, including those of temperate regions.[32,49] Following inundation, the dry cracked mud is actively eroded resulting in high water turbidity and the development of a substrate layer with a high proportion of organic carbon. During the wet phase, desert rainpools are often enriched by autochthonous (e.g., vegetation) and/or allochthonous (e.g., defecation by cattle) resources. In this environment of high productivity, the active population may be consistently below the carrying capacity.[17,32,50]

Under the above set of environmental conditions, population size (active + dormant component) will be mainly determined by the cumulative hatching percentage (both successful and abortive) at the end of the wet phase, by juvenile and adult mortality, by time of maturation, fecundity rate, duration of the wet phase of the pool, and by mortality rate and loss by dispersal in the dormant component. The resulting variability in long-term population size (typical for r-selected species), which is largely determined by the duration of the wet phase of the pool, is described by Southwood[24] as "boom and burst".

Apart from the adaptations to survive the dry phase, the capacities of desert rainpond species to maintain homeostasis in the fluctuating conditions (e.g. pH, O_2, conductivity, etc.) during the wet phase of the habitat also merit special attention. A discussion of these adaptations, however, would be beyond the scope of this chapter.

BIOLOGICAL AND LIFE CYCLE ADAPTATIONS

General

In spite of their extreme vulnerability to a variety of predators, freshwater branchiopods have successfully survived since the Devonian/Carboniferous, probably because of their specific adaptations which allow them to survive in extreme, but at the same time almost "enemy-free space"[51] provided by temporary pools.[51,52] Predator avoidance has indeed been suggested as one of the most important forces for the evolution of life-history strategies.[53,54] Reduced stress from biotic factors coupled with rich food resources are the rewards for breeding in ephemeral pools.[17]

The optimal life cycle strategy of an organism is to allocate its resources to maintenance, growth, and reproduction so as to maximize its contribution to the persistence of the population.[55,56] In particular, the reproductive strategies of inhabitants of temporary habitats must be adapted to compensate for adverse environmental conditions. For any organism, natural selection favors those traits that maximize its long term population size.[57]

Some qualitative and quantitative reproductive traits which maximize the benefits of life in temporary habitats of (semi-) arid regions will be discussed below, first with emphasis on their ecological significance (Figures 1, 2 and 3), then with regard to their merits for their exploitation in aquatic toxicology (Figures 1, 3, and 4).

Qualitative Reproductive Traits (Figure 1)

In fairy shrimps, tadpole shrimps, and clam shrimps, survival during adverse periods is ensured and their life cycles are closely geared to the cycle of drying and refilling by the production of shelled dormant embryos. Most branchiopods belong to the "life cycle type one" as distinguished by Wiggins et al.,[17] defined as permanent residents capable of passive dispersal only, and which avoid desiccation as resistant stages.

a. The extent to which dormancy (or diapause) in these animals is obligatory and controlled endogenously, or facultative and controlled exogenously is not always clear. In *S. proboscideus* and other freshwater branchiopods, dormancy is to some extent facultative (quiescence)

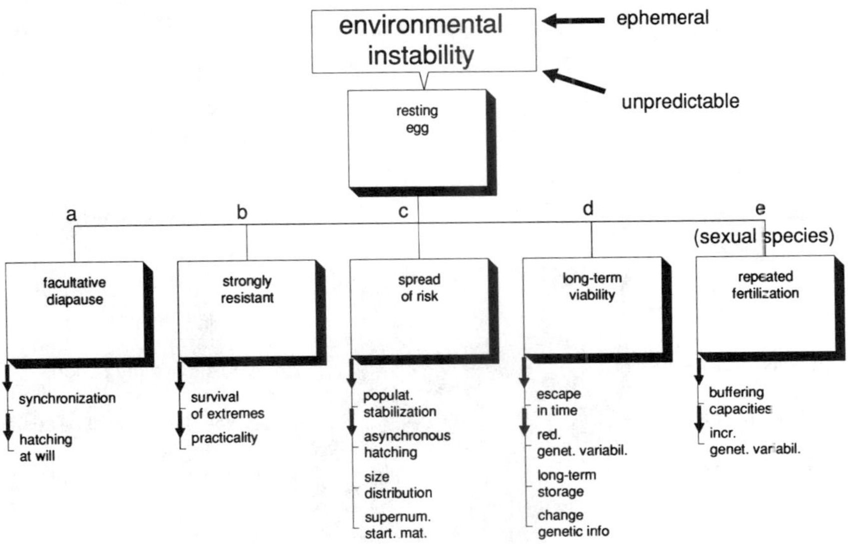

FIGURE 1. Diagram of qualitative reproductive traits of most freshwater branchiopods from (semi-) arid region, as adaptations to the unstable environment, their ecological significance (below first vertical arrow), and their consequences for exploitation in aquatic toxicology (below second vertical arrow).

(Figures 1a and 4), since the eggs hatch without a preceding dry period, by dilution or renewal of the culture medium.[30] In general, maximal population growth will occur when the embryos respond to those signals that have the highest correlation with subsequent growth and survival,[57] i.e., when the habitat becomes favorable again. With this strategy, the best synchronization of life cycle is obtained with the temporal heterogeneity of the habitat (Figures 1a and 2).[34, 58, 59]

Facultative diapausing or quiescent eggs offer major advantages for use in toxicity testing, since no elaborate diapause-deactivating techniques are required to obtain test-material[30] (Figures 1a and 4). These eggs can indeed be hatched at will by a simple rehydration or renewal of the medium.

b. Resting eggs are mainly distributed through the top 5 to 10 mm layer of the muddy sediment and are exposed there to harsh environmental conditions during the wet, and especially the dry phase of the pond. Resting eggs must therefore be very resistant (Figure 1b).

At the end of the wet phase, conductivity (due to evaporation) increases considerably. According to Beadle,[60] salinities of fresh temporary waters

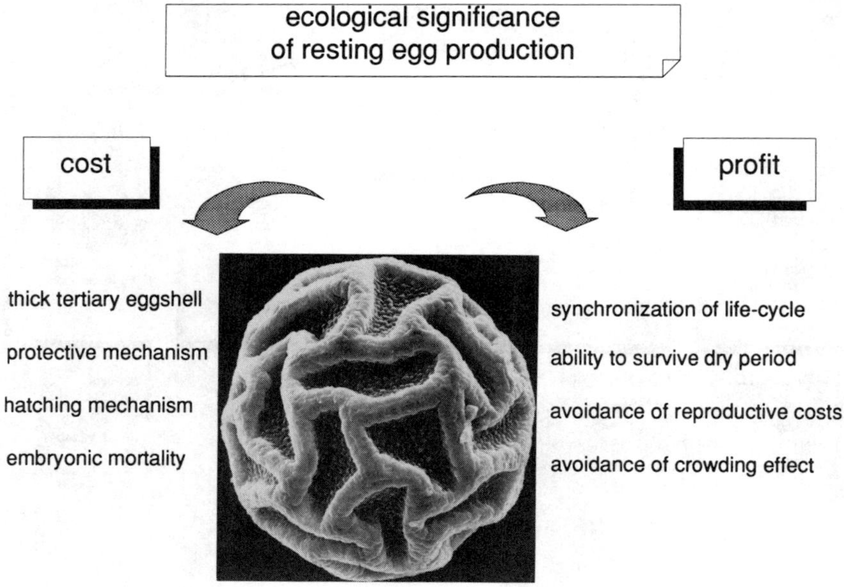

FIGURE 2. Diagrammatic presentation of the ecological significance (costs and profits) of resting egg production.

in the (semi-) arid zone can, upon drying, attain values inimical to life. Rzoska[40] reported a conductivity value of 1,130 μmhos. cm^{-1} (550 ppm) at the end of the existence of a temporary pond in the Sudanese desert. Brown and Carpelan[58] measured salinities of a Mohave desert (N-America) pond which were about equal to those of sea water. Eggs in the bottom mud, however, may be submitted to salinities that are even higher than those of the water layer above.[58]

The temperature in shallow ephemeral ponds closely follows air temperatures and strong fluctuations (both long-term seasonal and short-term diurnal) have often been recorded. At the surface of temporary pools, maximum temperatures ranging from 31.5°C up to 45°C have been measured.[36,40,61,62] Weir[38] measured maximum mud-temperatures during the wet-phase of Central African rainpools; these ranged from 25°C in the morning to 35°C in the afternoon. Hartland-Rowe[32] noted a temperature change rate of 4°C/h in a temporary pond in Alberta, Canada. *Streptocephalus seali* has been observed to occur from near-freezing temperatures in an ice-covered pool to sites with a maximum of 42°C.[41] Temperature fluctuations in the habitats of the Indian fairy shrimp species *Streptocephalus dichotomous* ranged from 20 to 46°C during daytime.[63] Dexter and Kuehnle[64] recorded temperatures from 0 to 21°C in the habitat of *Eubranchipus vernalis* and Nourisson[65] recorded a range of between 5 and 26°C for habitats of *Chirocephalus diaphanus* — both temperate species. The dry soil temperature of habitats of *Streptocephalus dichotomous* reached maxima from 47 to 72°C,[53] while Carlisle[66] and Cloudsley-Thompson[67] recorded the extremely high temperature of about 80°C in the dried mud from Sudanese ponds where *Streptocephalus proboscideus* eggs were sampled. Mitchell[68] measured in South African "pans" a maximum temperature of 66°C in mid-summer, at a depth of 2 mm in the dry sediment. During winter, ground frosts occurred nightly for about five months, with minimum temperatures reaching approximately –10°C during mid-winter.

Brown and Carpelan[58] found a sharp reduction in oxygen concentration just below soil/water interface with values of less than 1 mg/l at a depth of 5 to 10 mm in the mud. During the dry phase, heterotrophic respiration may cause a serious decrease in the oxygen levels at the surface, or just below the surface of the sediments.[36]

During the dry phase of the pond, eggs in the bottom sediments may also be exposed to very high light and U.V. radiation. Williams[36] recorded a daily amount of radiation input of about 24 MJ m^{-2} in Australian inland regions.

Moore[69] found the soil moisture of temporary ponds in Louisiana to range from 60% of the dry weight to as low as 9.5%. Sunbaked soils of dry pools in desert regions may have even lower water content for longer periods. Carlisle[66] could only kill the embryos of the tadpole shrimp *Triops granarius*, from Sudan, under temperature and pressure conditions that removed the last amount of water retained within the eggs. During strong dehydration, the embryos of *Limnadia stanleyana* (studied by Bishop[43]) and *Eulimnadia antlei* (studied by Belk[70]) shrunk considerably, and only occupied a small percentage of the total space available. The thick tertiary eggshell was shown to protect the embryo from mechanical injury (e.g., by sand grains) and from sunlight, but not from loss of water.[70]

The ability to withstand the above extremes of the physical environment (Figures 1b and 2) is a major advantage for storage, transportation, and manipulation of shelled dormant structures (Figure 1a).

c. Since, in freshwater branchiopods, colonization capacities in general, and dispersal powers in particular, are questionable,[44,71] survival in unstable environments with probability of total reproductive failure necessitates a combination of a high fecundity, a low yearly hatching fraction of resting stages, and the ability of "propagules" to survive for many generations in the soil to maintain a reserve of eggs to spread the risk.[72] A frequent threat of extinction of a Kenyan population of the fairy shrimp *Streptocephalus vitreus* is prevented by spread of the risk among the progeny[42] which is achieved by asynchronous hatching of the cysts (Figure 1c). This spread may result from micro habitat heterogeneity within a pond basin,[34] physical and chemical gradients from the soil/water interface downward,[43,58] or within-cyst clutch "balanced" polymorphisms in response to environmental cues. Similar distributions of hatching times have been found in seeds of many desert plants and in some mosquito (*Aedes*) eggs.[73] Species specific differences in requirements for embryonic development and egg hatching could, furthermore, explain year to year discontinuities in anostracan populations.[58,59,64,74–79] Hildrew[42] suggested that variation in behavior and diapause of *Streptocephalus vitreus* eggs can be inherited, a fact that also occurs in *Aedes* eggs.[80,81] Similarly, not all *Triops* eggs hatch at the same time even under favorable conditions.[82] Furthermore, the cysts of some species only hatch after repeated wet/dry cycles.[42] This "generation carry-over" by delayed development and prolonged diapause of a fraction of the population was found to stabilize and to increase the population density of a tadpole species, especially under unstable environmental conditions (Figure 1c).[83] According to Cohen,[72,84] only when the variance of the production of "propagules" of an annual organism becomes large in relation to the mean yield does it become advantageous to postpone (by dormancy) the hatching of some fraction of these structures. Consequently, the fraction that hatches yearly should be approximately equal to the probability of a large production, and the fraction which does not hatch should be almost equal to the probability of total or near total failure to produce "propagules".

Delayed and spread hatching are the opposites of synchronous hatching. The latter is sought in aquatic toxicology to obtain test animals of equal age and size, and a large amount of eggs is thus needed to obtain test organisms in sufficient numbers, at one particular moment, to start the bioassays (Figure 1c). In *Streptocephalus proboscideus*, for example, hatching of one batch of standard reference eggs is spread over more than one week (Figure 4).

d. Due to the unpredictability of rainfall in desert climates a high variability in the total production of resting eggs may be expected between different years. According to the optimal reproduction model of Cohen,[72,84] natural selection in such situations will tend to favor long-term viability of cysts (Figure 1d).

Organisms capable of producing resting eggs with long-term viability and spread of hatching are able to escape in time (Figure 1d), but at the same time avoid selection by environmental factors. This may finally result not only in retarded evolution,[85] but also in less fluctuations of genetic information than is the case in other biota (Figure 1d).[86] On the other hand, the simultaneous hatching of eggs from different generations (by spread of hatching) will, in turn, increase the genetic variability. Greenslade[87] suggested that adversity selection would favor conservation of adaptation in severe but predictable environments.

Reduced genetic variability would be favorable in toxicity testing because it would increase the uniformity of the test material, hence of the effects at the population level. Future research, however, is required to make conclusive statements.

High viability is a major advantage for long-term storage of cysts (Figure 1d). In the laboratory, some anostracan eggs have been shown to remain viable for up to 15 years.[88] Donald[74] found that eggs of *Branchinecta lindahli* remained unhatched and survived for a period of nine years under natural conditions. *Streptocephalus proboscideus* still hatched under laboratory conditions from the original resting egg-containing mud samples after a storage period of six years.

There are indications that in some animal or plant groups submitted to prolonged dehydration, replacement of messenger RNA occurred during the dry experience (Figure 1d).[89,90] It has not been fully determined yet if different durations of dry-storage of eggs of *S. proboscideus* have a significant effect on the sensitivity of the larvae.

e. In *S. proboscideus*, as in all fairy shrimps and the majority of clam shrimps, a female needs to be fertilized for each separate clutch of eggs[29] (Figure 1e). If there is some genetic basis for hatching variability, females producing successive clutches fertilized by different fathers might produce more variable progeny, which increases their fitness in an unpredictable environment.[42] It has been repeatedly stated that genetic diversity is intimately associated with ecological heterogeneity.[91,93] Nevo and Beiles[94] propose that species ranging across mesic and xeric environments display elevated levels of genetic diversity in hot deserts where climatic unpredictability is highest. Furthermore, they state that the level of genetic diversity may relate primarily to the buffering and homeostatic effect of heterozygosity (Figure 1e).

If increased heterozygosity is the case in large freshwater branchiopods it may cause a reduction in the uniformity of test material (Figures 1e and 4). The same remark, however, can be made for sexually produced resting stages in daphnids (with the exception of obligate parthenogenetic strains), rotifers, and *Artemia* (with the exception of some parthenogenetic strains), where meiotic divisions occur during gametogenesis.

f. It must be emphasized that the production of shelled eggs (Figure 2) is energetically more costly than viviparous reproduction. Furthermore, the synthesis of polyhydroxy compounds (such as trehalose) during the drying process to maintain the integrity of membranes during anhydrobiosis[95,96] and the perception and processing of external signals to hatch, are undoubtedly also energetically costly like many metabolic responses.[97] In addition, there is an unknown load of egg mortality both during the dry and the wet phase. The amount of resources allocated to such protective activities depends on the relation between the cost of such diversion vs. the benefit derived from it.[98] Rainpond species profit by the production of a thick tertiary egg shell around the dormant embryo as a protection from exposure to direct sunlight and blowing dirt during the dry season.

Quantitative Reproductive Traits (Figure 3)

The alternating cycle of favorable and unfavorable periods for reproduction maintains the ecosystem in an early successional stage (usually below the carrying capacity), with low competition and predation levels,[99] in which efficient colonizers will have an advantage.[17] The qualitative reproductive characteristics mentioned above have long been regarded as the keys to survival in temporary pools.[59,100] However, beside these, quantitative reproductive traits must also be considered.[27]

Based on the habitat templet of Southwood[24] relating the r-K continuum of MacArthur and Wilson[101] to the durational stability of the habitat, it can be predicted that success in discrete temporary desert pools will be favored by r-selection, which maximizes the intrinsic rate of natural increase, r, in a nonlimiting environment.[98,102–106] According to this theory, organisms in temporary waters imposing high density-independent mortality will be selectively favored to allocate a greater proportion of resources (time, matter, energy) to reproductive activities at the cost of growth and maintenance (Figure 3).[98] A review of

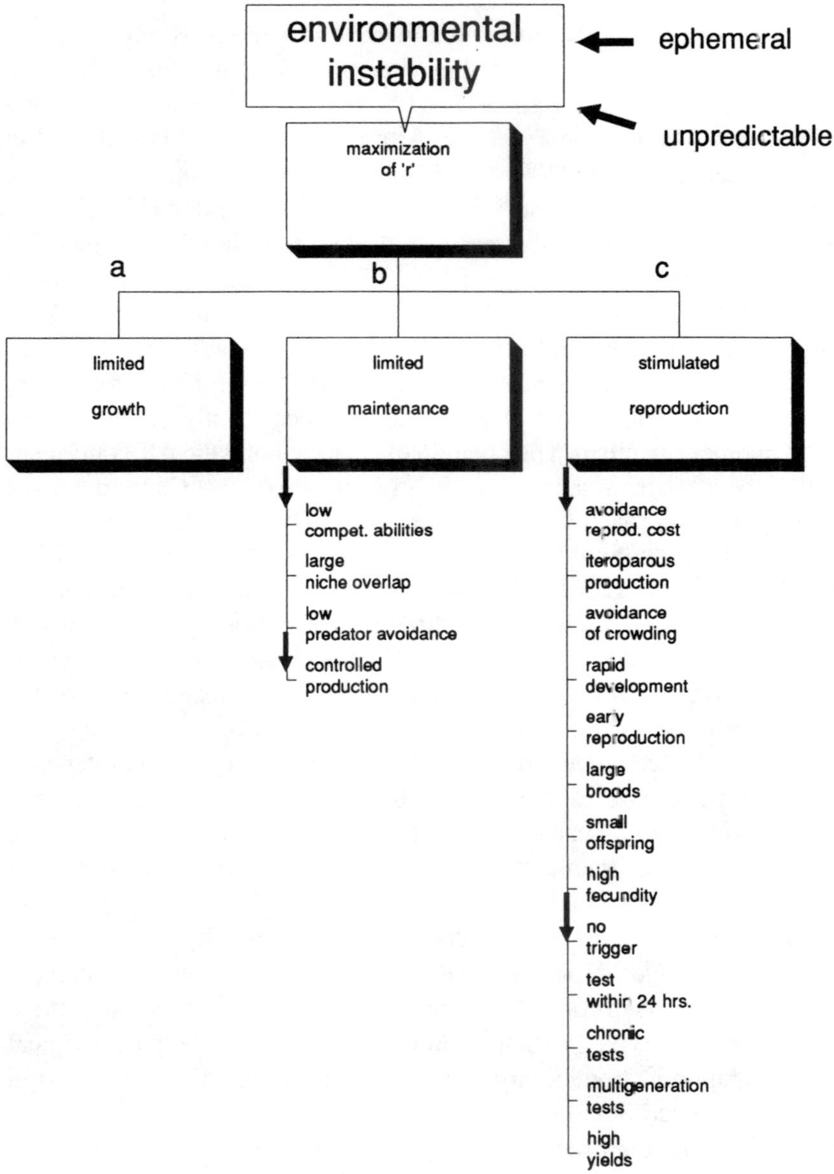

FIGURE 3. Diagram of quantitative reproductive traits of most freshwater branchiopods from (semi-) arid regions, as adaptations to the unstable environment, their ecological significance (below first vertical arrow), and their consequences for exploitation in aquatic toxicology (below second vertical arrow).

characteristics usually correlated with r-strategists is presented in Stearns.[106] It must be noted, however, that several other life cycle strategies have been used and every theory has its advocates and opponents. The r-K theory of MacArthur and Wilson[101] is still used in many recent works[17,36,86,107–109] and is also considered appropriate in the present case. While this sort of pattern (r-selection) does characterize many organisms of temporary waters and especially the organisms presently under discussion, there are, however, many departures from and variations upon it.[36] As Pianka[102,103] and others have stressed, no organism is entirely r- or K-selected. According to Gadgil and Solbrig,[98] the concepts of r- and K-selection are only meaningful by comparison; a given organism is more or less of an r-strategist only in comparison with another organism. The position of species along the r-K continuum will depend on the risk involved in any one situation.[17] Caution against a simplistic approach has been expressed by some authors.[110,111]

a. Reproduction is a heavy burden which cuts into growth,[55] resulting in slow growth after maturity (Figure 3a). In the absence of fish predation, size-selective feeding by invertebrate predators which appear in a later successional phase of the pond and specialize on small zooplankton,[56] will be the primary selective factor responsible for the relatively large size of temporary pond crustaceans.

b. A limited amount of energy is allocated to physical maintenance reducing the competitive abilities. Furthermore, since rainponds often have a high carrying capacity (see above), competition for food may be low. According to Pianka,[103] this low level of competition should result in a larger tolerable niche overlap (Figure 3b).

According to Takahashi,[82] tadpole shrimps generally do not have a characteristic such as self-regulation of population density to prevent over-exploitation of food. The organismic specialization enabling these organisms to survive in such rigorous habitats has led to an ecological generalization.[17] Productivity is the main objective, not efficiency as in more K-selected organisms.[98]

Furthermore, their reduced predator avoidance capacities make these organisms extremely vulnerable to fish and invertebrate predation (Figure 3b).

As a consequence of the supposed low level of competition in ephemeral pools, temporary pond inhabitants generally co-occur with several species. This creates an additional problem for the exploitation of resting eggs of freshwater branchiopods, since the eggs sink and a mixture of resting stages of various species settles in the sediments. In contrast, *Artemia* eggs float and can easily be harvested in a monospecific way. To obtain monospecific test material of freshwater branchiopods, it is therefore necessary to set up laboratory cultures of the organisms of interest (Figures 3b and 4). Controlled production of resting stages of a particular species, in turn, yields material of which the exact production conditions are known and thus offers opportunities to produce standard "reference" resting stages which are of relative constant quality.

A semi-automatic culture system for the controlled production of fairy shrimp resting stages as exemplified with *Streptocephalus proboscideus* is described in Brendonck et al.[20,21]

c. When adult mortality exceeds juvenile mortality (as is usually the case in r-selected species), theory predicts that organisms should reproduce only once in their life time (semelparity).[50,106,112] The probability of survival of subsequent active generations in an unpredictable environment indeed decreases in time, thus increasing the costs of reproduction. Any increment of fecundity would, without special survival strategy, be attained only at the expense of adding to the risk of mortality.[112] In the group of concern, however, which produces resting stages which fail to hatch before the next phase, the adverse effect on future survival is completely avoided and reproductive performance by iteroparous production of resting eggs is promoted. This high fecundity counterbalances for losses by hatching (successful and abortive), embryonal mortality, and dispersal (Figures 2 and 3c). Furthermore, Holgate[113] and Murphy[114] state that iteroparity — repeated breeding — will be directly selected when the risk of total reproductive failure in any given year is significant.

Furthermore, by producing a dormant next generation, most branchiopods avoid suppression of their growth rate caused by direct and indirect effects of crowding (increase in environmental resistance) (Figures 2 and 3c). These crowding effects generally reduce the growth potential of many plant and animal populations resulting in a different pattern with lower values than predicted with the logistic growth curve.[115]

Organisms inhabiting ephemeral ponds, and having only limited time to reach the reproductive stage before pools dry out or predators occur are literally involved in a 'race against time'. Consequently, time-efficient development and growth to sexual maturity are selected (Figure 3c). Rapid growth and reproduction has also been observed in several ponds formed from melting snow or runoff.[41,52,59,116] After inundation, there is a boom in the active population growth determined by the hatching response (fast/delayed). The rapid development after inundation may be augmented by protein-rich detritus derived from rapid decomposition during the previous dry period, leading to a considerable rise in aquatic production.[17,108] Rainfall starts a pulse of energetic biotic activity as predicted by the "pulse-reserve hypothesis".[117]

In expanding populations, selection will push the age for first reproduction to the physiological minimum. The age for reproduction will tend to increase as the availability of resources (the degree of satisfaction) decreases.[55] It seems reasonable to suppose that any delay in maturity can only add to the risk of dying before reproduction (increasing the reproductive cost). Moreover, the advantages of a younger age at first reproduction is greater for animals with large clutch sizes (Figure 3c). Most phyllopods are capable of a remarkably rapid development to sexual maturity with a relatively early start of reproduction at modest size.[29,34,40,42] *S. proboscideus* specimens cultured under laboratory conditions, for example, started egg production in less than one week and revealed the highest fecundity scores currently known for fairy shrimps[29,30] (Figure 4).

Generally, in rainponds, the optimal strategy in the absence of (both fish and invertebrate) predation on larval stages, is to put all possible matter and energy in reproduction, with the smallest practicable amount into each individual offspring, and to produce as many total progeny as possible (Figure 3c).[17,50,98,102,118] Belk[50] found, for fairy shrimps inhabiting rainponds, significantly larger clutches of smaller eggs than for species occurring in more predictable melt-ponds. The optimum reproductive effort increases with the availability of resources.[55]

The iteroparous and continuous production of resting eggs are appropriate for exploitation in aquatic toxicology and aquaculture, because no special physiological and environmental factors triggering gametogenesis and cyst formation need to be identified for the controlled production of resting eggs in most freshwater branchiopods in general and in *S. proboscideus* in particular (Figures 3c and 4). This contrasts with the vast amount of research done to identify the triggering factors for resting-egg formation in the rotifers *Brachionus plicatilis*, *B. rubens*, and *B. calyciflorus*. The literature on the induction of gametogenesis in daphnids is even more extensive and is reviewed in Persoone et al.[119] Although some cladoceran and rotifer species may also occur in ephemeral pools, the latter generally live in habitats which turn unfavorable with a certain periodicity, hence with higher predictability. It has been suggested that parthenogenesis provides a means of effective dispersal from the ancestral habitat but that retention of sexual

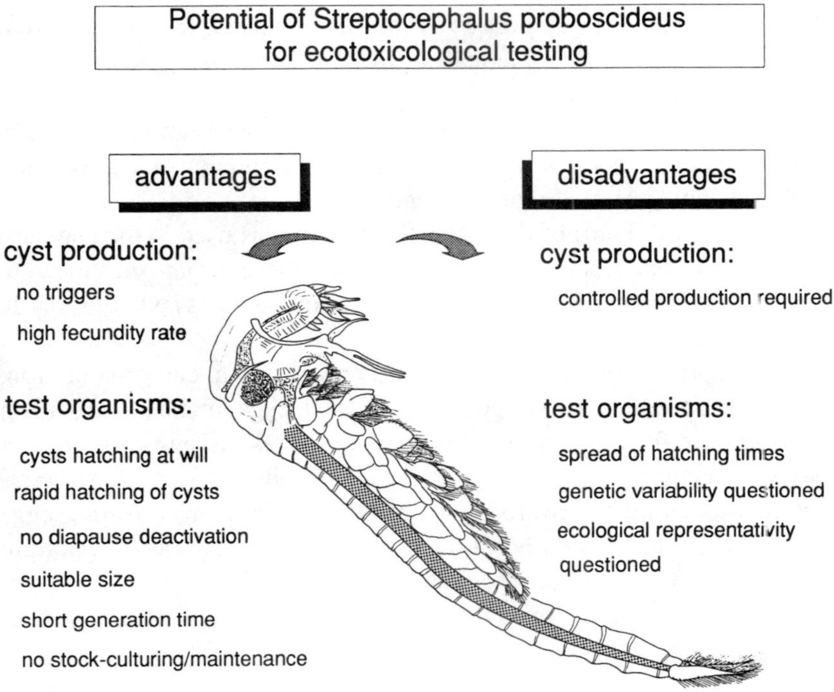

FIGURE 4. Diagrammatic presentation of advantages and disadvantages of the use of the fairy shrimp *Streptocephalus proboscideus* in aquatic toxicology.

reproduction ensures production of individuals best fitted to meet changing environmental conditions.[17] The resistant egg-envelopes in such organisms appear to have evolved in permanent waters where they protect the organism during unfavorable periods.[17] *S. proboscideus*, together with the other branchiopods from desert rainponds, on the other hand, cannot take the risk of first producing active offspring for some generations at reduced costs (no mating, no egg shell, and no dormancy process required), because early drying of the pond could lead to extinction of the active population and consequently to a considerable decrease of the total population size. Having a long geological history (dating back to Devonian periods), early evolution in non-cladoceran branchiopods was most probably influenced by very variable conditions.

In *S. proboscideus*, at 29°C, first hatching occurs 10 h after inundation. Such a quick hatching response enables harvesting of suitable test organisms (instar II-III), as soon as 24 h after hydration at 25°C (Figures

3c and 4).[23] Hatching response is very fast compared to the incubation period required to obtain neonates of *Daphnia magna* (3 d) and instar II-III larvae of *Artemia* (48 h).

The rapid growth, early maturation, and short generation time make these organisms also very attractive for development of chronic and multigeneration tests (Figures 3c and 4).

The size (±400 um) of the instar II-III larvae, as used in the standard *Streptocephalus* test[23] furthermore makes these organisms easy to manipulate, and imposes little bench space for the bioassays (Figures 3c and 4).

The adaptive reproductive traits (i.e., rapid growth, early maturation, iteroparous production of large broods of small eggs) make this organism very suitable for the controlled production of resting eggs, of which a large stock can be produced in a short time (Figures 3c and 4). In a semi-automatic culture system for the controlled production of resting eggs of *S. proboscideus*, a daily production of 18.000 eggs per 6 L-culture unit can easily be obtained at a culturing density of about 60 animals per liter, and increased production can be expected from further optimization.[21]

CRITIQUES

The major advantage of using resting stages as biological starting material for aquatic toxicity tests is the interesting alternative they offer to the problem of culturing or recruitment of live test organisms. A breakdown of the expenses into three major phases of toxicity testing (preparation, execution, and treatment of the data) revealed that the maintenance of live stock or the continuous culturing of the test organisms makes up a substantial proportion of the total (50% or more).[119,120] From a strictly financial point of view, the total costs of acute bioassays using rotifer cysts as starting material is 1 to 2 orders of magnitude more cost-effective for routine toxicity testing than those using stock-cultured organisms.[119,120] Since *Artemia* cysts are harvested in large amounts in nature and are commercially available, the costs of the biological material is practically negligible.

It can be expected that, like for some other cyst-based toxicity tests (see further), a major critique on the use of freshwater branchiopods from (semi-) arid rainpools, and of *S. proboscideus* in particular, as test species is their limited ecological relevance (in terms of representativity of the organisms supposed to occur in 'conventional' aquatic habitats and having there an important functional role). Occurring mainly in temporary ponds of sub-tropical African regions, it cannot be claimed that *S. proboscideus* is a major representative of the conventional freshwater ecosystem. Biogeographic data, however, indicate that the non-cladoceran freshwater branchiopods have the capacity to live in large permanent lakes. Their general exclusion from such environments is almost certainly a function of fish predation.[31] In the physical development of the Earth, freshwater temporary pools probably came into existence at least as early as permanent bodies of water and would thus have preceded the origin of life. The earliest inhabitants of temporary pools probably included ancestral branchiopods.[17,31] Currently, temporary waters are especially widespread in arid and semi-arid regions and have a much greater ecological significance there than in more temperate regions.[36]

As emphasized by Persoone,[121] the choice of a test organism will, however, always be a compromise between the factors "sensitivity" and "ecological representativeness" on the one hand, and "practicability" and "cost" on the other hand; each increase in ecological representativeness of the test will automatically lead to increased costs. It can likewise hardly be stated that *Artemia* is an important representative of the marine environment, since representatives of this genus only occur in salinas and hypersaline lakes and it cannot be forwarded either that *Daphnia magna* is a key-regulator of all freshwater ecosystems, since it is usually restricted to marginal hard water or polluted habitats[122] and shows a distinct preference for ponds (as does *D. pulex*).[123] However, the advantage of using resting stages as starting material makes this organism an extremely useful tool for screening studies of the relative toxicity of chemicals or effluents, or effects on population level responses such as growth or reproduction.

Finally, a critique on the use of cryptobiotic stages in general is that by using only a few species of cyst-producing organisms for ecotoxicological testing, such as those presently available, the battery of test species would not represent a random sample of species occurring in all types of biotops and belonging to all types of trophic levels. To increase the taxonomic diversity of test organisms, new tests may be developed based on cryptobiotic stages of micro-algae, gemmules of sponges, statoblasts of bryozoans, bacterial and fungal spores, resting copepodite stages of cyclopoid crustaceans, the larvae of some insects which can reversibly dehydrate,[124,125] organisms capable of anhydrobiosis during any stage of their life history such as some protozoans, nematodes and tardigrades, and even resistent eggs of fishes.

CONCLUSIONS

Due to the difficulties (technical, financial, etc.) in culturing or recruitment of healthy test organisms in sufficient numbers, there is not only a shortage of standardized aquatic bioassays but ecotoxicological testing is severely restricted for routine application.

The concept of using resting stages of various groups of organisms offers promising perspectives for the development of a battery of cost-effective toxicity tests to complement the bioassays presently used.

In contrast to more commonly used resting-egg producing organisms (the brine shrimp *Artemia*, the daphnids *Daphnia magna* and *D. pulex*, and the rotifers *Brachionus plicatilis*, *B. calyciflorus* and *B. rubens*), which, disregarding some exceptional strains, alternate an asexual (viviparous) and sexual (oviparous) life-cycle phase, most freshwater non–cladoceran branchiopods continuously produce resting eggs in a sexual (fairy shrimps) or asexual way (some clam shrimps and tadpole shrimps) and thus offer a promising and still unexploited source for the development of resting stage-based tests.

In particular, the unique combination of adaptive life cycle traits as presented by some rainpool species of arid or semi-arid regions (e.g., strongly resistant resting eggs, high fecundity rate, rapid growth, early maturation, rapid hatching, small size, and short generation time) makes these organisms attractive for applications not only in aquatic toxicology but also in aquaculture, as live foods for rearing larval and juvenile stages of species of commercial interest.

To obtain monospecific test material of constant and high quality, however, a culture system for the controlled production of resting stages needed to be developed, and has been worked out.

The practicality of using 'standard reference' resting eggs which can be hatched at will must be weighted against the limited ecological relevance of rainpool species, the limited choice of resting egg producing organisms, the possible genetic variability in sexually produced resting eggs, and the variability in hatching response of the eggs. The many advantages definitely outweigh the limitations, and make non-cladoceran freshwater anostracans very appealing candidates for routine toxicity testing of hazardous products.

Streptocephalus proboscideus, a subtropical fairy shrimp species possessing the capacity for a very high fecundity, can to date be successfully cultured under controlled conditions. The development of a standard short-term toxicity test with eggs of this species, is presented in a second paper.[23]

ACKNOWLEDGMENTS

The senior author is Research Assistant with the National Fund for Scientific Research (NFWO, Belgium). The authors thank Dr. K. Martens and Dr. De Meester for their constructive comments.

REFERENCES

1. Persoone, G. and P.G. Wells. 1987. *Artemia* in aquatic toxicology: a review. In P. Sorgeloos, D.A. Bengtson, W. Decleir and E. Jaspers, Eds., *Artemia Research and Its Applications*. Universa Press, Wetteren, Belgium, pp. 259–275.
2. Snell, T.W. and G. Persoone. 1989a. Acute toxicity bioassays using rotifers. I. A test for brackish and marine environments with *Brachionus plicatilis*. *Aquat. Toxicol.* 14: 751–765.
3. Snell, T.W. and G. Persoone. 1989a. Acute toxicity bioassays using rotifers. II. A freshwater test with *Brachionus rubens*. *Aquat. Toxicol.* 14: 81–92.
4. Van Steertegem, M. and G. Persoone. This volume. Development and critical evaluation of standardized toxicity tests with the brine shrimp *Artemia* (Crustacea: Anostraca).
5. Peters, N. Jr. 1963. Embryonale Anpassungen oviparer Zahnkarpfen aus periodisch austrocknenden Gewässern. *Intern. Rev. Hydrobiol.* 48: 257–313.
6. Wourms, J.P. 1972. The developmental biology of annual fishes. III. Pre-embryonic and embryonic diapause of variable duration in eggs of annual fishes. *J. Exp. Zool.* 182: 389–414.
7. Bogaert, T., G. Persoone and D. Dive. 1982. Research on the development of a short term bioassay for water and soil, with amoebae, starting from cysts. *ISERM* 106: 565.

8. Dive, D. and G. Persoone. 1984. Protozoa as test organisms in marine ecotoxicology: luxury or necessity? In G. Persoone, E. Jaspers and C. Claus, Eds., *Ecotoxicological Testing for the Marine Environment*, Vol. I, Univ. Ghent and Ins. Mar. Scient. Res., Bredene, Belgium, pp. 281–305.
9. Halbach, U., M. Wiebert, C. Wissel, J. Klaus, K. Beuter and M. Delion. 1981. Population dynamics of rotifers as bioassay tools for toxic effects of organic pollutants. *Verh. Inter. Verein. Limnol.* 21: 1141–1146.
10. Halbach, U., M. Wiebert, M. Westermayer and C. Wissel. 1983. Population ecology of rotifers as a bioassay tool for ecotoxicological tests for aquatic environments. *Ecotox. Envir. Safety* 7: 484–513.
11. Snell, T. W., B.D. Moffat, C.R., Janssen and G. Persoone. Acute toxicity bioassays using rotifers: IV. Effect of age, temperature and salinity on the sensitivity of *Brachionus calyciflorus. Aquat. toxicol.* (in press).
12. Snell, T. W., B.D. Moffat, C.R., Janssen and G. Persoone. Acute toxicity bioassays using rotifers: III. Effects of temperature, strain and exposure time on the sensitivity of *Branchionus calyciforus. Aquat. Toxicol.* (in press).
13. Adema, D.M.M. 1978. *Daphnia magna* as a test animal in acute and chronic toxicity tests. *Hydrobiologia* 59:125–134.
14. Leeuwangh, P. 1978. Toxicity tests with daphnids: its application in the management of water quality. *Hydrobiologia* 59:145–148.
15. Ten Berge, W.F. 1978. Breeding *Daphnia magna. Hydrobiologia* 59:121–123.
16. Buikema, A.L. Jr., J.G. Geiger and D.R. Lee. 1980. *Daphnia* toxicity tests. In A.L. Buikema, Jr. and J. Cairns, Jr., Eds., *Aquatic Invertebrate Bioassays*, ASTM STP 715, American Society for Testing and Materials, pp. 48–69.
17. Wiggins, G.B., R.J. Mackay and I.M. Smith. 1980. Evolutionary and ecological strategies of animals in annual temporary pools. *Arch. Hydrobiol., Suppl.* 58:97–206.
18. Brendonck, L. and G. Persoone. 1990a. Preliminary results concerning the culture of freshwater anostracans and conchostracans (Crustacea: Branchiopoda) for applications in aquaculture and aquatic toxicology. *Belg. J. Zool.* 120:219.
19. Brendonck, L. and G. Persoone. 1990b. The culture of freshwater anostracans and conchostracans (Crustacea: Branchiopoda) for applications in aquaculture and environmental sanitation. *Belg. J. Zool.* Suppl. 120, 1:11–12.
20. Brendonck, L., G. Uyttersprot and G. Persoone. 1990. A culture system for fairy shrimps (Crustacea: Anostraca). *Belg. J. Zool.* 120:220.

21. Brendonck, L., G. Uyttersprot and G. Persoone. 1991. A culture system for fairy shrimps (Crustacea: Anostraca). *Aquacultural Engineering* 9:267–283.

22. Centeno, M.D., L. Brendonck and G. Persoone. 1990. The development of a short-term standard toxicity test with the fairy shrimp *Streptocephalus proboscideus* (Crustacea: Anostraca). *Belg. J. Zool.* Suppl.120, 1:13–14.

23. Centeno, M.D., L. Brendonck and G. Persoone. This volume. Cyst-based toxicity tests. III. Development and standardization of an acute toxicity test with the fairy shrimp *Streptocephalus proboscideus* (Crustacea: Branchiopoda: Anostraca).

24. Southwood, T.R.E. 1977. Habitat, the templet for ecological strategies? *J. Anim. Ecol.* 46:337–365.

25. Brendonck, L. 1990a. Redescription of the fairy shrimp *Streptocephalus proboscideus* (Frauenfeld, 1873) (Crustacea: Anostraca). *Bulletin Van het Koninklijk Belgisch Instituut voor Natuurwetenschappen, Biologie* 59(1989):49–57.

26. Brendonck, L. 1990b. Contributions to the study on the feeding biology of the fairy shrimp *Streptocephalus proboscideus* (Crustacea: Branchiopoda: Anostraca). *Belg. J. Zool.* 120:10–11.

27. Brendonck, L. in press. Feeding in the fairy shrimp *Streptocephalus proboscideus* (Frauenfeld)(Branchiopoda: Anostraca). I. Aspects of the feeding biology. *J. Crust. Biol.* 13.

28. Brendonck, L. in press. Feeding in the fairy shrimp *Streptocephalus proboscideus* (Frauenfeld)(Branchiopoda: Anostraca). II. Influence of environmental conditions on feeding rate. *J. Crust. Biol.* 13.

29. Brendonck, L. 1991. Contributions to the study of the reproductive biology of *Streptocephalus proboscideus* (Anostraca, Streptocephalidae). *Crustaceana* 60:145–162.

30. Brendonck, L., D.M. Centeno, and G. Persoone. in press. Fecundity and resting egg characteristics of some subtropical fairy shrimp and clam shrimp species (Crustaces: Branchiopoda) reared under laboratory conditions. *Arch. Hydrobiol.*

31. Kerfoot, W.C. and M. Lynch. 1987. Branchiopod communities: associations with planktivorous fish in space and time. In W.C. Kerfoot and As. H, Eds., *Predation, Direct and Indirect Impacts on Aquatic Communities.* University Press, New England, Hanover, pp. 367–378.

32. Hartland-Rowe, R. 1972. The limnology of temporary waters and the ecology of Euphyllopoda. In R.B. Clark and R. Wooton, Eds., *Essays in Hydrobiology.* University of Exeter, U.K., pp. 15–31.

33. Fryer, G. 1987. A new classification of the branchiopod Crustacea. *Zool. J. Linn. Soc.* 91:357–383.

34. Belk, D. and G.A. Cole. 1975. Adaptational biology of desert temporary-pond inhabitants. In N.F. Hadley, Ed., *Environmental Physiology of Desert Organisms*. Dowden, Hutchinson and Ross, Inc., Stroudsburg, PA, pp. 207–226.
35. Decksbach, N.K. Von. 1929. Zur Klassifikation der Gewasser vom astatischen Typus. *Arch. Hydrobiol.* 20:399–406.
36. Williams, W.D. 1985. Biotic adaptations in temporary lentic waters, with special reference to those in semi-arid and arid regions. *Hydrobiologia* 125:85–110.
37. Morgan, N.C. and Boy, V. 1982. An ecological survey of standing waters in north west Africa: I. Rapid survey and classification. *Biological Conservation* 24:5–44.
38. Weir, J.S. 1969. Studies on Central African Pans: III. Fauna and physico-chemical environment of some ephemeral pools. *Hydrobiologia* 33:93–116.
39. Cantrell, M.A. and Mc Lachlan, A.J. 1982. Habitat duration and dipteran larvae in tropical rain pools. *Oikos* 38:343–348.
40. Rzoska, J. 1961. Observations on tropical rain pools and general remarks on temporary waters. *Hydrobiologia* 17:265–297.
41. Moore, W.G. 1955. Observations on heat death in the fairy shrimp *Streptocephalus seali*. *Proc. Louisiana Acad. Sci.* 18:5–12.
42. Hildrew, A.G. 1985. A quantitative study of the life history of a fairy shrimp (Branchiopoda: Anostraca) in relation to the temporary nature of its habitat, a Kenyan rainpool. *J. Anim. Ecol.* 54:99–110.
43. Bishop, J.A. 1968. Resistance of *Limnadia stanleyana* King (Branchiopoda, Conchostraca) to desiccation. *Crustaceana* 14:35–38.
44. Bishop, J.A. 1967. Some adaptations of *Limnadia stanleyana* King (Crustacea: Branchiopoda: Conchostraca) to a temporary freshwater environment. *J. Anim. Ecol.* 36:599–609.
45. Bishop, J.A. 1967. Seasonal occurrence of a branchiopod crustacean, *Limnadia stanleyana* King (Conchostraca) in eastern Australia. *J. Anim. Ecol.* 36:77–95.
46. Sublette, J.E. and M.S. Sublette. 1967. The limnology of playa lakes on the Llando Estacado, New Mexico and Texas. *Southwest. Nat.* 12:369–406.
47. Louw, G.N. and M.K. Seely. 1982. *Ecology of Desert Organisms*. Longman, NY, pp. 1–194.
48. Alderice, D.F. 1972. Responses of marine poikilotherms to environmental factors acting in concert. In O. Kinne, Ed., *Marine Ecology*. Wiley-Interscience, U.K., pp. 1659–1722.
49. Mc Lachlan, A.J. 1981. Food sources and foraging tactics in tropical rain pools. *Zool. J. Linn. Soc.* 71:265–277.

50. Belk, D. 1977. Evolution of egg size strategies in fairy shrimps *Southwest. Nat.* 22:99–105.
51. Fryer, G. 1986. Enemy-free space: a new name for an ancient ecological concept. *Biol. J. Linn. Soc.* 27:287–292.
52. Ardö, P. 1948. Some notes on phyllopods in temporary pools on the Alvar of Öland in South Sweden. *K. Fys. Sällsk. Handl. N.F.*59:1–22.
53. Jefferies, M.J. and J.H. Lawton. 1984. Enemy free space and the structure of ecological communities. *Biol. J. Linn. Soc.* 23:269–286.
54. Maier, G.U. 1989. Variable life cycles in the freshwater copepod *Cyclops vicinus* (Uljamin 1875): support for the predator avoidance hypothesis? *Arch. Hydrobiol.* 115:203–219.
55. Gadgil, M. and W.H. Bossert. 1970. Life historical consequences of natural selection. *Am. Nat.* 104:1–24.
56. Lynch, M. 1977. Fitness and optimal body size in zooplankton populations. *Ecology* 58:763–774.
57. Cohen, D. 1967. Optimizing reproduction in a randomly varying environment when a correlation may exist between the conditions at the time a choice has to be made and the subsequent outcome. *J. Theoret. Biol.* 16:1–14.
58. Brown, L.R. and L.H. Carpelan. 1971. Egg hatching and life history of a fairy shrimp *Branchinecta mackini* Dexter (Crustacea: Anostraca) in a Mohave Desert playa (Rabbit Dry Lake). *Ecology* 52:41–54.
59. Broch, E.S. 1965. Mechanism of adaptation of the fairy shrimp Chirocephalopsis bundyi Forbes to the temporary pond. *Cornell Univ. Agr. Exp. Sta. Mem.* 392:1–48.
60. Beadle, L.C. 1981. The inland waters of tropical Africa. An introduction to tropical Limnology, 2nd ed., Longman, U.K.
61. Hillyard, S.D. and A. Vinegar. 1972. Respiration and thermal tolerance of the phyllopod Crustacea *Triops longicaudatus* and *Thamnocephalus plattyurus* inhabiting desert ephemeral ponds. *Physiol. Zool.* 45:189–195.
62. Otto, R.G. and S.D. Gerking. 1973. Heat tolerance of a theath valley pupfish (Genus Cyprinodon). *Physiol. Zool.* 46:43–49.
63. Bernice, R. 1972. Ecological studies on *Streptocephalus dichotomous* Baird. *Hydrobiologia* 39:217–240.
64. Dexter, R.W. and C.H. Kuehnle. 1948. Fairy shrimp populations of Northeastern Ohio in the seasons of 1945 and 1946. *Ohio J. Sci.* 48:15–26.
65. Nourisson, M. 1964. Récherches écologiques et biologiques sur le Crustacé Branchiopod *chirocephalus diaphanus* Prév. Etude experimentelle du déterminisme du développement de l'oeuf. Docteur ès Science Naturelles Thèse, Université de Nancy, France.

66. Carlisle, D.B. 1968. *Triops* (entomostraca) eggs killed only by boiling. *Science* 161.279.
67. Cloudsley-thompson, J.l. 1965. The lethal temperature of Triops granarius (Lucas) (Branchiopoda: Notostraca). 25:424–425.
68. Mitchell, S.A. 1990. Factors affecting the hatching of *Streptocephalus macrourus* Daday (Crustacea; Eubranchiopoda) eggs. *Hydrobiologia* 194:13–22.
69. Moore, W.G. 1967. Factors affecting egg-hatching in *Streptocephalus seali* (Branchiopoda, Anostraca). *Proc. Symp. Crustacea, Marine Biol. Assoc.* Part 2. India, pp. 724–735.
70. Belk, D. 1970. Functions of the conchostracan egg shell. *Crustaceana* 19:105–106.
71. Brendonck, L., A. Thiery, and A. Coomans. 1990. Taxonomy and biogeography of the Galapagos branchiopoda fauna (Anostraca, Notostraca, Spinicaudata). *J. Crustacean Biol.,* 10:676–694.
72. Cohen, D. 1966. Optimizing reproduction in a randomly varying environment. *J. Theoret. Biol.* 12.119–129.
73. Gillet, J.D. 1959. Control of hatching in the prediapause eggs of *Aedes* mosquitoes. *Nature* 184:1621–1623.
74. Donald, D.B. 1983. Erratic occurrence of anostracans in a temporary pond. Colonization and extintion or adaptation to variations in annual weather. *Canad. J. Zool.* 61:1492–1498.
75. Dexter, R.W. 1946. Further studies on the life history and distribution of *Eubranchipus vernalis* (Verrill). *Ohio J. Sci.* 46:31–44.
76. Dexter, R.W. 1953. Studies on North American fairy shrimps with the description of two new species. *Am. Midl. Nat.* 49:751–771.
77. Prophet, C.W. 1963. Egg production by laboratory-cultured Anostraca. *Southwest. Nat.* 8:32–37.
78. Horne, F. 1967. Effects of physical-chemical factors on the distribution and occurrence of some southeastern Wyoming phyllopods. *Ecology* 48:472–477.
79. Horne, F.R. 1971. Some effects of temperature and oxygen concentrations on phyllopod ecology. *Ecology* 52:343–347.
80. Gillet, J.D. 1955a. Variation in the hatching-response of *Aedes* eggs (Diptera: Culicidae). *Bull. Ent. Res.* 46:241–254.
81. Gillet, J.D. 1955b. The inherited basis of variation in the hatching-response of *Aedes* eggs (Diptera: Culicidae). *Bull. Ent. Res.* 46:255–265.
82. Takahashi, F. 1977a. Pioneer life of the tadpole shrimps, *Triops* spp. (Notostraca: Triopsidae). *Appl. Ent. Zool.* 12:104–117.
83. Takahashi, F. 1977b. Generation carryover of a fraction of population members as an animal adaptation to unstable environmental conditions. *Res. Popul. Ecol.* 18:235–242.

84. Cohen, D. 1968. A general model of optimal reproduction in a randomly varying environment. *Ecology* 56:219–228.
85. Hairston, N.G. Jr. and De Stasio Jr. 1988. Rate of evolution slowed by a dormant propagule pool. *Nature* 336:239–242.
86. Freckman, D.W. 1986. The ecology of dehydration in soil organisms. In A.C. Leopold, Ed., *Membranes, Metabolism, and Dry Organisms.* Comstock Publishing Associates, Cornell University Press, Ithaca, NY, pp. 157–168.
87. Greenslade, P.J.M. 1983. Adversity selection and the habitat templet. *Am. Nat.* 122:352–365.
88. Moore, W. G. 1979. The effect of prolonged wet storage on the hatchability of eggs of *Streptocephalus seali* Ryder (Branchiopoda: Anostraca). *Southwest. Nat.* 24:201–203.
89. Leopold, A.C. 1986. Introduction. In A.C. Leopold, Ed., *Membranes, Metabolism, and Dry Organisms.* Comstock Publishing Associates, Cornell University Press, Ithaca, NY, pp. 17–21.
90. Kermrode, A.R. and J.D. Bewley. 1986. Alteration of genetically regulated syntheses in seeds by desiccation. In A.C. Leopold, Ed., *Membranes, Metabolism, and Dry Organisms.* Comstock Publishing Associates, Cornell University Press, Ithaca, NY, pp. 59–84.
91. Nevo, E. 1983. Population genetics and ecology: the interface. In D.S. Bendall, Ed., *Evolution from Molecules to Men.* Cambridge University Press, Cambridge, U.K., pp. 287–321.
92. Nevo, E., A. Beiles and R. Ben-shlomo. 1984. The evolutionary significance of genetic diversity: ecological, demographic and life history correlates. In G.S. Mani, Ed., *Evolutionary Dynamics of Genetic Diversity.* Lecture notes in Biomathematics 53:13–213.
93. Nevo, E. and A. Beiles. 1988. Genetic parallelism of protein polymorphism in nature: ecological test of the neutral theory of molecular evolution. *Biol. J. Linn. Soc.* 35:229–245.
94. Nevo, E. and A. Beiles. 1989. Genetic diversity in desert: patterns and testable hypotheses. *J. Arid Env.* 17:241–224.
95. Crowe, J.H., L.M. Crowe and D. Chapman. 1984. Preservation of membranes in anhydrobiotic organisms: the role of trehalose. *Science* 223:701–703.
96. Crowe, J.H. and L. M. Crowe. 1986. Stabilization of membranes in anhydrobiotic organisms. In A.C. Leopold, Ed., *Membranes, Metabolism, and Dry Organisms.* Comstock Publishing Associates, Cornell University Press, Ithaca, NY, pp. 188–209.
97. Young, B.A., B. Walker, A.E. Dixon and V. A. Walker. 1989. Physiological adaptation to the environment. *J. Anim. Sci.* 67:2426–2432.
98. Gadgil, M. and O.T. Solbrig. 1972. The concept of r- and K-selection: evidence from wild flowers and some theoretical considerations. *Am. Nat.* 106:14–31.

99. Varon, R. and D. Thistle. 1988. Response of a harpacticoid copepod to a small-scale natural disturbance. *J. Exp. Mar. Biol. Ecol.* 118:245–256.
100. Longhurst, A.R. 1955. Evolution in the Notostraca. *Evolution* 9:84–86.
101. MacArthur, R.H. and E.O.Wilson. 1967. *The Theory of Island Biogeography*. Princeton University Press, Princeton, NJ.
102. Pianka, E.R. 1970. On r- and K-selection. *Am. Nat.* 104:529–597.
103. Pianka, E.R. 1972. r- and K-selection or b and d selection? *Am. Nat.* 106:581–588.
104. Southwood, T.R.E., R.M. May, M.P. Hassell and G.R. Conway. 1974. Ecological strategies and population parameters. *Am. Nat.* 108:791–804.
105. Allan, J.D. 1976. Life history patterns in zooplankton. *Am. Nat.* 110:165–180.
106. Stearns, S.C. 1976. Life history tactics: a review of the ideas. *Q. Rev. Biol.* 51:3–47.
107. Calow, P. 1981. Adaptational aspects of growth and reproduction in *Lymnaea peregra* (Gastropoda: Pulmonata) from exposed and sheltered aquatic habitats. *Malacologia* 21:5–13.
108. Williams, D.D. 1987. *The Ecology of Temporary Waters*. Croom Helm, London, Timber Press, Portland, OR.
109. Mc Lachlan, A.J. 1983. Life-history tactics of rain-pool dwellers. *J. Anim. Ecol.* 52:545–561.
110. Wilbur, H.M., D.W. Tinkle and J.P. Collins. 1974. Environmental certainty, trophic level, and resource availability in life history evolution. *Am. Nat.* 108:805–817.
111. Yom-Tov, Y. 1989. Environmental predictability and life history traits in arid environments. *J. Arid Env.* 17:193–197.
112. Bell, G. 1980. The costs of reproduction and their consequences. *Am. Nat.* 116:45–76.
113. Holgate, P. 1967. Population survival and life history phenomena. *J. Theor. Biol.* 14:1–10.
114. Murphy, G.I. 1968. Pattern in life history and the environment. *Am. Nat.* 102:390–404.
115. Hall, C.A.S. 1988. An assessment of several of the historically most influential theoretical models used in ecology and of the data provided in their support. *Ecological Modelling* 43:5–31.
116. Johansen, F. 1921. The larger freshwater Crustacea from Canada and Alaska. *Can. Field Nat.* 35:21–30.
117. Noy-Meir, I. 1973. Desert ecosystems: environment and procedures. *Ann. Rev. Ecol. Syst.* 4:25–52.
118. Dodson, S.I. 1974. Zooplankton competition and predation: an experimental test of the size efficiency hypothesis. *Ecology* 55:605–613.

119. Persoone, G., L. Brendonck, E. Cotou, C. Janssen, A. Van De Vel and M. Van Steertegem. 1990. EEC 6614/89/44, Final report, Commission of the European Communities, Ghent, Belgium.

120. Persoone, G. and A. Van de Vel. 1987. Cost-analysis of 5 current aquatic ecotoxicological tests. EUR 1134EN. Environment and quality of life. Commission and quality of life. Commission of the European Communities. DGXI-Environment, Consumer Protection and Nuclear Safety, Ghent, Belgium.

121. Persoone, G. 1980. Standardization of aquatic bioassays: compromises between biological and economical criteria. *Can. Tech. Rep. Aquatic. Sci.* 975:111–122.

122. Flossner, D. 1972. *Die Tierwelt Deutschands. 60. Krebstiere, Crustacea. Kiemen-und Bladfüsser, Branchiopoda. Fischläuse, Branchiura.* V. E. B. Gustav Fisher Verlag, Jena.

123. Fryer, G. 1985. Crustacean diversity in relation to the size of water bodies: some facts and problems. *Freshwater Biol.,* 15:347–361.

124. Hiton, H.E. 1960. Cryptobiosis in the larva of *Polypedilum vanderplankii* Hint. (Chironomidae). *J. Insect Physiol.* 5:286–300.

125. Lees, A.D. 1961. Insect diapause in relation to the environment. In N. Grossowicz, S. Hestrin and A. Reyman, Eds., *Cryptobiotic Stages in Biological Systems.* Elsevier, NY, pp. 120–131.

126. Calleja, M.C. and G. Persoone. 1992. Cyst-based toxicity tests. IV. The potential of ecotoxicological tests for the prediction of acute toxicity in man as evaluated on the first ten chemicals of the M.E.I.C. programme. *Atla* 20:396–405.

CHAPTER 3

Cyst-Based Toxicity Tests. III. Development and Standardization of an Acute Toxicity Test with the Freshwater Anostracan Crustacean *Streptocephalus proboscideus*

Maria Dulce Centeno, Luc Brendonck, and Guido Persoone

ABSTRACT

A new method is presented for determination of the acute toxicity of chemicals and effluents to larvae of the freshwater anostracan *Streptocephalus proboscideus* (Frauenfeld), using resting eggs (cysts) as the source of the test organisms. The larvae are obtained by controlled hatching of standard reference eggs. For the testing laboratory, this procedure eliminates all the difficulties inherent in recruitment or continuously culturing live stocks.

0-87371-845-3/93/$0.00 + $0.50

Appropriate test conditions (temperature, dilution medium, photoperiod, and duration) have been selected and a protocol has been developed for the execution of a 24-h toxicity test, with scoring of larval mortality as the effect criterion. Standard Reference Water (SRW) is used as hatching and dilution medium for the bioassay, which is carried out in polystyrene multiwell plates, at 25°C, in darkness. Since the sensitivity of the larvae changes with age, instar II-III nauplii obtained within 24 h from the start of hatching are used for the test.

24-h LC_{50}s determined with the new test procedure on 8 chemical compounds gave the following ranking of decreasing sensitivity (mg/L) in: $HgCl_2$ (0.03) — $CuSO_4$ (0.21) — $CdCl_2$ (0.46) — NaPCP (0.80) — trichloroacetic acid (1.2) — $K_2Cr_2O_7$ (1.9) — $ZnSO_4$ (4.0) — sodium lauryl sulfate (23).

The *Streptocephalus proboscideus* acute toxicity test is a useful addition to the present limited battery of cyst-based toxicity tests (Toxkits). The sensitivity and reliability of the Streptoxkit test compare favorably with those of other commonly used aquatic invertebrate bioassays.

KEYWORDS

Acute toxicity test, Anostraca, *Streptocephalus proboscideus*, resting stages, bioassay method.

INTRODUCTION

The scarcity of test species which can be recruited or cultured in good condition and in sufficient numbers for routine aquatic toxicity testing is acknowledged worldwide by ecotoxicologists. Toxicity testing on a variety of species of remote phylogenetic relationships, reasonably representative for aquatic biota is, in general, the only way to determine the "overall" range of the hazardous nature of a compound.[1]

The majority of the standard freshwater toxicity tests used to date are based on a few species of microalgae, cladoceran crustacean, and fish. In recent years, interest has grown in the use of freshwater anostracans as additions to the limited battery of tests currently available. However, only a few studies have been published dealing with the effects of toxic compounds on fairy shrimps. Moss[2] reports the toxicity of selected chemicals under laboratory and field conditions on adult stages of the species *Streptocephalus seali*. The 48-h LC_{50} of the insecticide phosmet was also determined for *S. seali* using mature specimens.[3,4] The biological and technological difficulties, as well as financial constrains associated with continuous culturing and maintenance of live test organisms[5] has, to date, probably been the major factor limiting the use of fairy shrimps for testing.

However, fairy shrimps have similar characteristics to the brine shrimp *Artemia salina*[6–8] and the rotifers *Brachionus plicatilis*, *Brachionus rubens*,[9,10] and *Brachionus calyciflorus*[11] which make them attractive candidates for aquatic toxicity test species. Like their halobiontic counterpart *Artemia*, freshwater anostracans are capable of producing resting eggs (commonly called "cysts"), which can be dry-stored for long periods of time without losing their viability thus bypassing recruitment and culturing problems associated with maintaining stock cultures. *Streptocephalus proboscideus*, a subtropical anostracan species occurring in Sudan and southern Africa, has a high fecundity, a short generation time, and reproduces exclusively through production of resting (cryptobiotic) eggs.[12,13] A culture system designed by Brendonck et al.[14] allows the production of resting eggs which can be stored and then subsequently used in toxicity tests.

This paper describes a new acute toxicity test with *Streptocephalus proboscideus* based upon larvae hatched from cysts produced under controlled conditions. This test will be referred to as the "Streptoxkit F" test—an analogy to the three other cyst-based Toxkit tests developed recently in the laboratory of the authors, with the halobiontic anostracan crustacean *Artemia* and the freshwater and estuarine rotifers *Brachionus* (Artoxkit M, Rotoxkit F, Rotoxkit M). The major biotic and abiotic parameters, as well as the technical factors involved in Streptoxkit F test, have been screened and fixed for the standard operational procedure of this new test. The susceptibility of the new test species has been determined for a number of compounds, and has been compared to results published for other aquatic organisms currently in use in acute toxicity testing.

Since the term "cysts" is generally established through the wide application of *Artemia* eggs in aquaculture and toxicology, the terms "cysts" and "resting eggs" shall be used interchangeably below.

MATERIALS AND METHODS

Cyst Production, Processing, and Hatching

Cysts of *Streptocephalus proboscideus* were originally collected from dry temporary pools in Al Gedid, near Khartoum in Sudan.[12] The larvae which hatched from these cysts were cultured to the adult stage under controlled laboratory conditions in a semi-automatic culture system.[14] The cysts produced by the latter animals were subsequently harvested, cleaned (washing and removal of debris), and dry-stored under defined conditions. All test animals referred to in this study originate from laboratory-produced cysts.

Hatching is initiated by rehydrating dried cysts with suitable synthetic freshwater (SRW or EPA medium, see further) at a density of 25 mg cysts in 100 ml medium, in a cylindrico-conical glass tube placed near a light source (1000–2000 lux). Incubation is temperature-controlled at 25 ± 1°C and the cysts are kept in suspension by gentle bottom aeration (at a rate of 1 or 2 bubbles per second). After 16 to 18 h of incubation, all free-swimming larvae are harvested by gentle suction with a glass pipet (0.5 cm diameter) and transferred to an Erlenmeyer flask containing fresh medium for a period of 5 to 8 h to allow molting into instar II-III stage. The latter step, which separates early-hatching larvae from delayed hatchers, allows for uniformity in age of the test animals. As indicated below, tests have been carried out with larvae of various ages to select the instar stage most appropriate for a standard test.

Determination of the Influence of Dilution Medium, Temperature, Light Regime, and Test Duration on Larval Survival

To select suitable test conditions, experiments have been carried out to determine the influence of dilution medium composition, temperature, light regime, and test duration on the survival of *Streptocephalus proboscideus* larvae. Three artificial freshwaters were considered for use as dilution media for the toxicants: Standard Reference Water (SRW),[15] EEC water,[16] and EPA water.[17] SRW and EPA water are both moderately hard synthetic freshwaters (hardness of 80 to 100 mg/L as $CaCO_3$), with a pH of 7.9 ± 0.2 and 7.6 ± 0.2, respectively. EEC water has a higher hardness (250 ± 25 mg/L $CaCO_3$) and a pH of 7.6 ± 0.2. Of these respective test media, 1 ml was distributed to each well of 24-well polystyrene plates. Nauplii in the instar II-III stage were subsequently transferred, with a glass micropipette with smoothed openings, into multiwells containing SRW, EPA, and EEC water, respectively, at a density of 10 animals per well. The plates were incubated at two different temperatures (20°C and 25°C), and at three photoperiods: 24 h light (24L/0D), 12 h light and 12 h darkness (12L/12D), and 24 h darkness (0L/24D), respectively. The experimental set-up, therefore, was a 3 × 2 × 3 medium-temperature-photoperiod factorial design carried out in six replicates. Mortality of the larvae was assessed after 24, 48, and 72 h.

Toxicity Test Procedure

The experimental testing procedure is based on the principles and recommendations for ecotoxicological testing outlined in "Standard Methods for the Examination of Water and Wastewater",[18] with slight modifications based on considerations of simplicity and practicality. The test procedure is analogous to that developed for the cyst-based toxicity tests with *Artemia* larvae (Artoxkit M).

The static test is conducted in a conventional 24-well polystyrene multiwell plate (6 columns with 4 rows). Each test comprises one control and five toxicant concentrations, with three duplicates per concentration. Each replicate consists of one well, receiving ten nauplii. As for the *Artemia* test,[6] the selection of three parallels with ten test organisms per concentration is a compromise between test precision and economic considerations (such as time, materials, etc.).

The plates are loaded using the procedure described in Figure 1. The 4 wells of the first column (column 1) are filled with 1 ml of the selected dilution medium; those of the next 5 columns (columns 2 to 6) each received 1 ml of the respective toxicant solutions, in the order of increasing concentrations. The nauplii are first transferred, with the aid of a micropipette, into the 6 wells of the bottom row (D), which serve as rinsing wells. The rinsing well of the control column receives approximately 35 nauplii which are then subsequently transferred into the 3 wells (A to C) of the same column (10 animals per well). This operation, which is carried out under a dissection microscope (10 to 12x

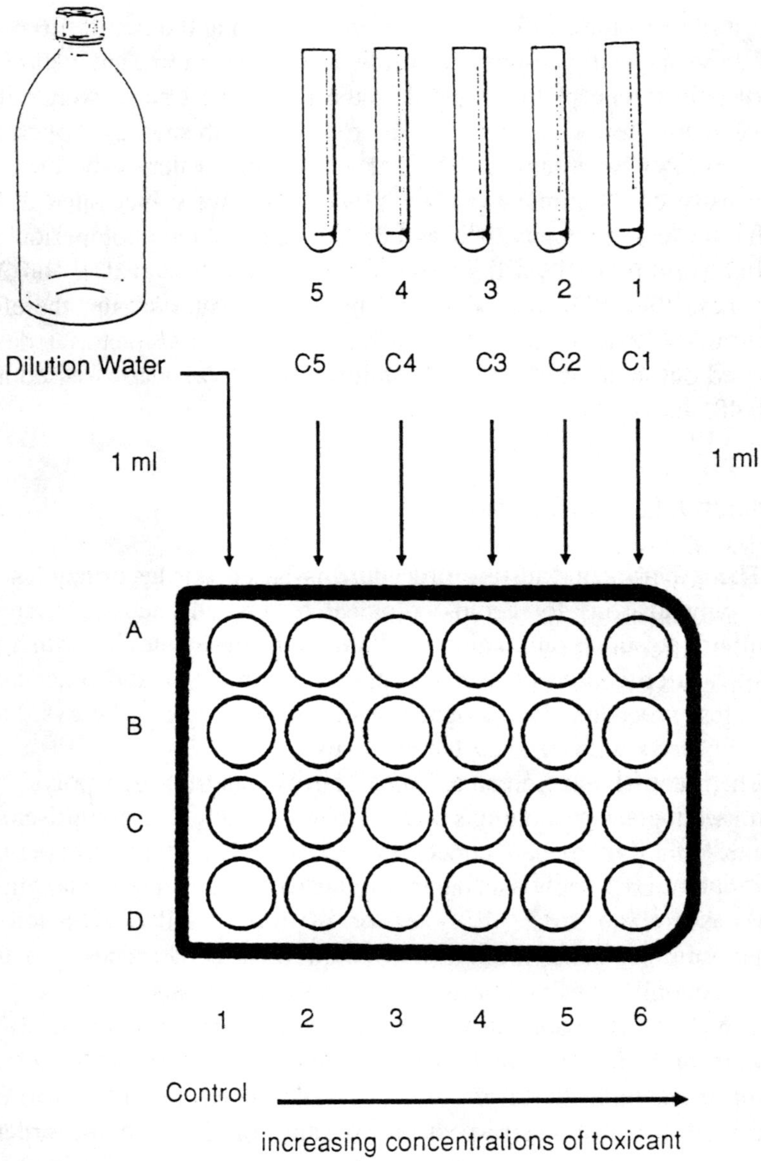

FIGURE 1. Diagram showing the sequence in which the wells are filled in a test plate. (C5 = lowest toxicant concentration; C1 = highest toxicant concentration.)

magnification), is subsequently repeated for each column (2 to 6), progressing from low to high toxicant concentrations.

The intermediate passage of the larvae through the rinsing wells "washes" them in the respective test solutions before they enter the actual test well. This minimizes dilution of the test solutions during the transfer process.

To avoid evaporation of the test medium during incubation, the multiwells are covered with a Parafilm strip prior to closing with the matching polystyrene lid. The plates are incubated in darkness at 20°C and 25°C for periods of 6 and 24 h. The specific conditions are outlined hereafter in the different experiments performed for selection of appropriate test conditions.

On the termination of the exposure, the number of dead larvae in each well is counted under a dissection microscope. Larvae are considered to be dead when no movement of the appendages is noted within 10 s of observation.

The 24-h LC_{50} and the 95% confidence limits are calculated following the Moving Average Method.[19] The slope functions are normally calculated following the Probit method, although in a few cases (when partial kills are less than 2) the Trimmed Spearman-Karber method was applied.[20] Differences between variables were analyzed with the one- and multiple-way ANOVA.

Influence of Instar Stage, Duration of Exposure, and Temperature on Larval Sensitivity

To determine which instar stage was most sensitive and most appropriate for the test (control mortality not more than 10%), hatched larvae were harvested 16 h after incubation of the dried cysts (the first larvae appeared after about 14 to 15 h). Half of this "instar I" larval population was used immediately for the test on instar-I larvae, while the other half was transferred in an Erlenmeyer flask containing 100 ml of gently aerated SRW medium, for further incubation at 25°C. Regular checks of the developmental stage revealed that approximately 5 to 8 h after the transfer, more than 80% of the nauplii had molted into the second or even the third instar stage. These larvae were then collected to perform bioassays on the instar II-III nauplii. Copper sulfate has been used as the reference toxicant in these experiments. After preliminary range-finding experiments, the following $CuSO_4$ concentrations were selected for the definitive tests: 0.056 — 0.10 — 0.18 — 0.32 — 0.56 — 1.0 mg/L.

For selection of a suitable test duration, results were scored 6 h and 24 h after the start of the test. For selection of a suitable temperature, the test organisms were incubated at 20 and 25°C. All tests were carried out in darkness and there were ten replicates for each treatment.

Determination of the Sensitivity of Early- vs. Late-Hatching Nauplii

To investigate differences in sensitivity of instar II-III nauplii hatching from the cysts after different periods of incubation, instar I-larvae were harvested sequentially. Eighteen hours after the start of cyst incubation, hatched nauplii were collected for the first time and transferred to a flask containing 100 ml SRW medium. These nauplii were incubated for a further 6 h, at 25°C, with illumination and gentle aeration to allow molting into instar II-III larvae. Twenty-four-hour toxicity tests were then performed on these nauplii using copper sulfate as a reference toxicant. The tests were performed in SRW medium and at 25°C in darkness, with three replicates. The same procedure was followed for larvae which hatched later and which were collected respectively after 20, 24, 26, 28, and 30 h after the start of cysts incubation.

Sensitivity of Instar II-III Nauplii to Various Chemicals in Two Different Dilution Media

Eight chemicals commonly used as reference compounds in ecotoxicology have been assayed to identify which would be suitable for use as a reference chemical for quality control in a routine test procedure. The selected analytical-grade chemicals tested were copper sulfate ($CuSO_4.5H_2O$, AG, UCB), sodium pentachlorophenate (98% NaPCP, Merck), potassium dichromate (99% $K_2Cr_2O_7$, Fluka), cadmium chloride ($CdCl_2.2.5H_2O$, AG, UCB), sodium lauryl sulfate (99% SDS, Merck), trichloroacetic acid (99% TCA, Riedel-De Haen), mercuric chloride (99.5% $HgCl_2$, Farmitalia Carlo Erba), and zinc sulfate (99.5% $ZnSO_4.7H_2O$, Ferak Lab). All tests were conducted on instar II-III nauplii, in SRW dilution medium, at 25°C in darkness for 24 h. The tests were run at least 9 times for each chemical.

RESULTS

Influence of Dilution Medium, Temperature, Light Regime, and Test Duration on Larval Survival

The cumulative percentages of larval mortality as a function of time in the three media used and in the different temperature and light regimens are shown in Table 1. The results reveal that in both SRW and EPA Water, larval mortality was less than 10% after 24 h in all tested conditions. In the EEC medium, substantial mortality was noted after 24 h in 5 out of 6 tested conditions. Only at 20°C, and in continuous light did more than 90% of the larval population survive for 24 h in the EEC medium.

Extension of the incubation period to 48 h did not result in more than 10% mortalities in both SRW and EPA medium at a temperature of 20°C. At 25°C, however, the percentage mortality increased up to 13% in the

Table 1
Cumulative Percentage Mortality of Larvae as a Function of Time, in Different Media at Different Temperatures and Photoperiods

Medium	Temperature (°C)	Light regime	Mean cumulative percentage mortality with standard deviation					
			24 h		48 h		72 h	
			x	±SD	x	±SD	x	±SD
SRW	20°C	24L/0D	0	—	1.6	4.1	5	8.4
		12L/12D	1.7	4.1	1.7	4.1	10	11
		0L/24D	1.7	4.1	1.7	4.1	20	16
	25°C	24L/0D	1.7	4.1	13	15	85	16
		12L/12D	0	—	12	8	80	13
		0L/24D	0	—	10	9	82	17
EEC	20°C	24L/0D	2.2	2.7	43	13	52	8
		12L/12D	27	12	45	10	58	10
		0L/24D	28	12	52	18	57	10
	25°C	24L/0D	45	8.4	54	10	97	5.2
		12L/12D	38	15	60	6.3	98	4.1
		0L/24D	53	10	62	17	98	4.1
EPA	20°C	24L/0D	0	—	0	—	0	—
		12L/12D	3.4	8.2	3.4	8.2	13	8.2
		0L/24D	1.7	4.1	5	8.4	20	11
	25°C	24L/0D	0	—	0	—	67	12
		12L/12D	0	—	3.3	5.2	70	24
		0L/24D	3.3	5.2	8.3	12	93	8.2

SRW medium, at all light regimes. In the EEC medium, larval mortality was substantial (40 to 60%) after 48 h, irrespective of the light regime.

After 72 h incubation, mortality percentages were below 10% only in the SRW and EPA media at an incubation temperature 20°C and 24 h exposure to light. At 25°C, mortality was very high in all media and under all conditions.

Influence of Instar Stage, Duration of Exposure, and Temperature on Larval Sensitivity

The differences in sensitivity to copper sulfate of larval stage I vs. stages II and III are shown in Table 2 (as LC_{50}s) for the various experiments carried out at 20 and 25°C, and for 6 h and 24 h of exposure to the toxicant. At the 6-h test interval, statistical analysis of the data revealed that instar II-III are significantly more sensitive than instar I nauplii ($p < 0.05$), although the difference is only a factor of 2. Prolonging the exposure period from 6 to 24 h decreases the LC_{50}s. Increase of the incubation temperature from 20 to 25°C, however, does not influence the LC_{50}s significantly ($p < 0.05$).

Sensitivity of Early- vs. Late-Hatching Nauplii to Copper Sulfate

The comparison of the sensitivity to copper sulfate of the early- vs. the late-hatching nauplii is shown in Table 3. The data clearly indicate that there is no difference in sensitivity of instar II-III to the compound used, for larvae hatching within 18 to 28 h after the start of the incubation of the cysts; 24-h LC_{50}s only ranged from 0.19 to 0.23 mg/L. For late-hatchers (30 h after the start of cyst incubation), a marginal increase in LC_{50} was found (24-h LC_{50} = 0.33 mg/L), but was still of the same order of magnitude.

Table 2
Toxicity of Copper Sulfate to Instar I vs. Instar II-III Larvae, Incubated in Darkness at 20 and 25°C

Instar stage	Temperature (°C)	6-h LC50	24-h LC50
I	20°C	0.70 (0.64–0.76)	0.35 (0.29–0.41)
I	25°C	0.66 (0.59–0.73)	0.29 (0.20–0.38)
II–III	20°C	0.36 (0.34–0.38)	0.23 (0.20–0.26)
II-III	25°C	0.35 (0.31–0.39)	0.21 (0.18–0.24)

Note: Mean 6- and 24-h LC50s with confidence limits; n = 10

Table 3
Toxicity of Copper Sulfate to Nauplii Harvested at Different Times After the Start of Incubation of the Cysts

Time of harvest (hours after start of cyst incubation)	24-h LC50
18	0.21 (0.17–0.28)
20	0.20 (0.17–0.25)
24	0.19 (0.16–0.23)
26	0.21 (0.18–0.24)
28	0.23 (0.19–0.27)
30	0.33 (0.27–0.42)

Note: Mean 24-h LC50s with 95% confidence limits; n = 10

Susceptibility of Instar II-III Larvae to Various Toxicants in SRW and EPA Media

The 24-h LC_{50}s obtained with the 8 compounds in the 2 dilution media are given in Table 4. The 24-h LC_{50} values determined for the reference chemicals range from 0.033 mg/L (for $HgCl_2$ in SRW) to 29 mg/L (for SDS in EPA medium). The results indicate that with the exception of sodium lauryl sulfate (SDS), instar II-III larvae are very sensitive to the chemicals assayed even at low concentrations.

The toxicity tests also revealed that, depending on the compound, the sensitivity of the larvae to the toxicant could differ by a factor of 2 depending upon the dilution medium used for the test. However, the differences are chemical-dependent, in the sense that for one compound the 24-h LC_{50} is lower in SRW than in EPA medium, whereas for the other compounds it is higher. Finally, the 95% confidence limits of the replicates indicate that the precision of the bioassay is quite satisfactory; the range of variation coefficients is generally only from 7 to 21%, with only one outlier of 29%.

Table 4
Toxicity of Various Compounds to Instar II-III Larvae, on Exposure at 25°C, in Darkness, in SRW vs. EPA Dilution Medium

Chemical	24-h LC50	In SRW	24-h LC50	In EPA
$CuSO_4$	0.21	(0.18–0.24)	0.15	(0.13–0.17)
$HgCl_2$	0.033	(0.028–0.040)	0.064	(0.060–0.070)
$CdCl_2$	0.46	(0.42–0.50)	0.51	(0.49–0.53)
NaPCP	0.80	(0.75–0.85)	0.87	(0.87–0.90)
$K_2Cr_2O_7$	1.9	(2.0–2.6)	0.89	(0.83–0.95)
$ZnSO_4$	4.0	(3.7–4.2)	3.0	(2.9–3.0)
SDS	23	(19–27)	29	(28–30)
TCA	1.2	(1.0–1.3)	1.5	(1.3–1.6)

Note: Mean 24-h LC50s with confidence limits; n = 10

DISCUSSION

Cyst Hatching and Molting of the Larvae

Most cysts of anostracans from temporary pool environments do not readily hatch[21] and even under carefully controlled conditions of incubation, hatching of a batch of cysts is not synchronous.[6] For *Streptocephalus proboscideus*, hatching extends over a period of several days and is influenced by physical as well as by chemical conditions of hatching.[13] It is, therefore, imperative that the hatching procedures and conditions described in the standard operational procedure are strictly adhered to in order to obtain test organisms of the same age and the same physiological state. *Streptocephalus proboscideus* larvae which are produced according to the procedures outlined above provide test animals (instar II-III) ready for bioassays within 24 h. For the halobiontic anostracan crustacean *Artemia salina*, instar II-III nauplii can only be obtained 48 h after the start of cyst rehydration. This is probably due to the slower molting rate of this crustacean.[6]

Dilution Media, Temperature, Light Regime, and Test Duration

The results of the various tests performed on survival of the nauplii in media without toxicants indicate that both SRW and EPA are suitable synthetic freshwaters for an acute toxicity test with *Streptocephalus* larvae. However, as shown in Table 4, the LC_{50}s can differ in the two dilution media depending on the chemical compound tested. Preference is given to the EPA medium for the standard acute test with *Streptocephalus proboscideus*, both as hatching and dilution medium since this artificial freshwater is easier to prepare than SRW.

Taking into consideration that both test duration and temperature have an influence on larval mortality (through metabolic activity), the combination of 24-h exposure and 25°C incubation temperature seems a good compromise. 25°C was selected as the incubation temperature to be consistent with the temperature required for cyst hatching and larval molting. The selection of a 24-h exposure period was dictated by the substantial gain in sensitivity when compared to the results obtained with a 6-h exposure period (Table 2).

Since the light regime, in the preferred combination of test duration and incubation temperature, does not affect larval survival, preference is given to total darkness (0L/24D) even when testing non-photodegradable compounds.

Toxicity Test Procedure

From the various experiments carried out, it appears that from the technological point of view, the "multiwell plate technology" developed for the *Artemia* acute bioassay (Artoxkit M) is very satisfactory. According to the data generated in the replicated tests, the precision of the bioassay technique used is very good. Variation coefficients indeed were in most cases below 20%.

Selection of *Streptocephalus* Naupliar Stage for the Acute Toxicity Test

The outcome of the test with copper sulfate as toxicant with instar I nauplii vs. instar II-III nauplii confirm the findings of Vanhaecke et al.[6–8] for *Artemia*, namely that the older larvae (i.e., instar II-III) are more sensitive. Consequently, as in the Artoxkit M test, the procedure for the standard acute toxicity test with *Streptocephalus* nauplii will prescribe the use of instar II-III larvae. The results presented in Table 3 show that the sensitivity of the instar II-III larvae does not vary between "early-hatching" nauplii, and nauplii hatching 10 h later.

Sensitivity of the *Streptocephalus* Nauplii Acute Test — Selection of a Reference Chemical

The results of the determination of the acute toxicity of the various chemicals to *Streptocephalus* larvae (Table 4) indicated that this test organism, like many others, is sensitive to some compounds, in particular the heavy metal mercury (24-h LC_{50} = 0.064 mg/L in EPA medium), but not as much as to others such as sodium lauryl sulfate (24-h LC_{50} = 29 mg/L in EPA medium)

In view of the fact that *Streptocephalus* nauplii have also been shown to be sensitive to potassium dichromate (24-h LC_{50} in EPA medium = 0.89 mg/L), and that this compound is frequently used in aquatic toxicology as a reference chemical, it is proposed that it should be used as a reference standard for this test.

Table 5 compares the susceptibility of the *Streptocephalus* nauplii to 4 compounds under the described test conditions with that of 8 other species tested in accordance with standard procedures. From this table, it appears that for the 4 compounds, the 24-h LC_{50}s for the *Streptocephalus* larvae are, generally, of the same order of magnitude as the acute LC_{50}s obtained for some of the other species, freshwater as well as estuarine-marine. However, with some exceptions *Streptocephalus* was found to be 46 times less susceptible than the Microtox™ bacteria (*Photobacterium phosphoreum*) to sodium lauryl sulfate, but is almost 100 times more susceptible than *Artemia salina* to copper, and 170 times more susceptible than *Brachionus plicatilis* to cadmium.

Bearing in mind that toxicity is species-specific, chemical-specific, and condition-specific, it can be concluded that the sensitivity of this new cyst-based toxicity test compares favorably with that of other standard tests.

Table 5
Comparison of the Relative Susceptibility of *Streptocephalus* Nauplii (Instar II-III) to Four Chemicals with that of Eight Other Acute Toxicity Test Species

Chemicals		SP	BC	DM	DP	MB	AS	BP	MT	FHM
Copper	x	0.050	0.020[a]	0.050[24]	0.050[24]	0.18[25]	4.9[b]	0.075[a]	0.40[26]	0.43[27]
	cv	19%	11%	—	—	—	—	4.4%	—	—
Cadmium	x	0.23	1.3[a]	0.13[22]	0.48[22]	0.11[25]	—	39[a]	8.0[26]	0.43[27]
	cv	11%	12%	82%	45%	—	—	4.2%	—	—
NaPCP	x	0.80	1.2[a]	0.50[22]	0.52[22]	—	0.51[b]	1.9[a]	0.60[26]	0.21[4]
	cv	7.0%	3.9%	32%	34%	—	—	9.2%	—	—
SDS	x	23	1.3[a]	17[22]	15[22]	7–19[23]	19[b]	5.6[a]	0.50[26]	8.7[17]
	cv	22%	7.9%	40%	36%	—	—	9.8%	—	—

Note: Mean LC50s in mg/L, with coefficient of variation. SP = *Streptocephalus proboscideus;* BC = *Brachionus calyciflorus;* DM = *Daphnia magna;* DP = *Daphnia pulex;* MB = *Mysidopsis bahia;* AS = *Artemia salia;* BP = *Brachionus plicatilis;* MT = Microtox™; FHM = fathead minnow *Pimephales promelas.* All LC50s are 24-h tests, except for MB (96 h), MT (30 min EC50) and FHM (96 h).

[a] C. Janssen (pers. communication) State University of Ghent
[b] M. Van Steertegem (pers. communication) State University of Ghent

CONCLUSIONS

In view of the need for cost-effective acute toxicity tests for routine monitoring of wastes and effluents, research has been performed with the freshwater anostracan *Streptocephalus proboscideus*. A standard acute toxicity test was developed using instar II-III nauplii of this species as test organisms, hatched from cysts. The implications of the results of this research for defining test parameters were as follows:

- Several artificial freshwaters have been compared from which the SRW and/or the EPA media have finally been selected as appropriate hatching and dilution media for an acute test.
- Standard hatching conditions for the cysts and molting conditions for the larvae were defined. Tests with nauplii in the first vs. the second and third instar indicated that the latter age group was more susceptible to toxicants, and easier to obtain as a group of uniform age and physiological condition.
- Potassium dichromate ($K_2Cr_2O_7$) was selected as a convenient reference chemical for quality control tests.
- The precision of the new test is quite satisfactory; the variation coefficient for repeated tests is indeed in most cases below 20%.
- The sensitivity of the cyst-based *Streptocephalus* nauplii acute toxicity test compares favorably with that of standardized acute toxicity tests with other species.

Figure 2 is a schematic diagram of the steps involved in the preparation and execution of the *Streptocephalus* test, following the standard test protocol.

The *Streptocephalus* nauplii test has, in common with three other cyst-based acute toxicity tests (*Artemia salina*, *Brachionus calyciflorus*, and *Brachionus plicatilis*), been tailored into a TOXKIT procedure that is now available commercially* under the name STREPTOXKIT F.

*Manufacturing and distribution: Creasel Ltd., Fazantenpark 9, 9800 Deinze, Belgium

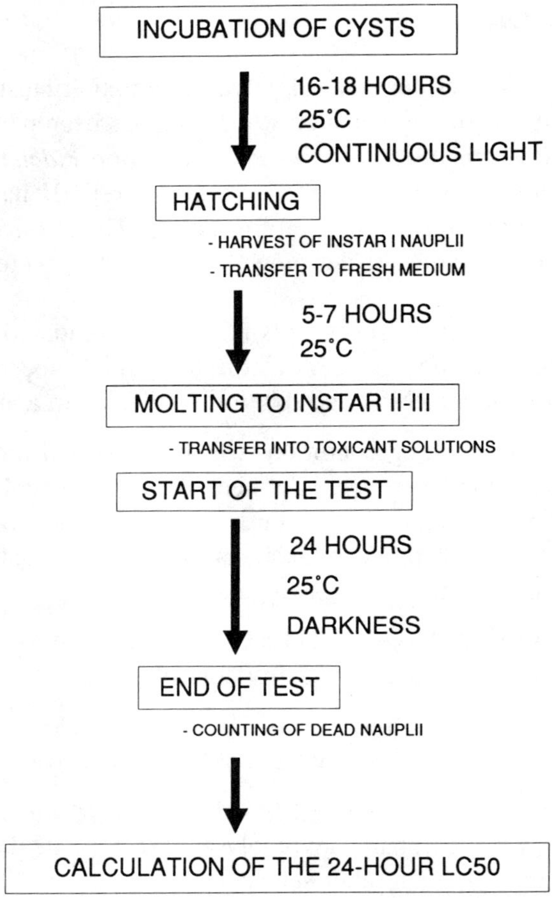

FIGURE 2. Schematic diagram and time schedule for preparation and execution of an acute toxicity test with *Streptocephalus proboscideus* nauplii.

ACKNOWLEDGMENTS

This research was supported by a grant from the Belgian Administration for International Cooperation (BAIC). L.B. is Research Assistant with the National Fund for Scientific Research (NFWO, Belgium).

REFERENCES

1. Persoone, G., D. Calamari and P. Wells. 1990. Possibilities and limitations of predictions from short-term tests in the aquatic environment. In P. Bourdeau, et al., Eds., *Short-Term Toxicity Tests for Non-Genotoxic Effects*, SCOPE 41. John Wiley and Sons, Ltd., Chichester, pp. 301–310, Chapt. 18.
2. Moss, J.L. 1978. Toxicity of selected chemicals to the fairy shrimp, *Streptocephalusseali*, under laboratory and field conditions. *Prog. Fish-Cult.* 40:158–160.
3. Julin, A.M. and H.O. Sanders. 1977. Toxicity and accumulation of the insecticide Imidan in freshwater invertebrates and fishes. *Trans. Am. Fish. Soc.* 106:386–392.
4. Johnson, W.W. and M.T. Finley. 1980. *Handbook of Acute Toxicity of Chemicals to Fish and Aquatic Invertebrates.* Resource Publication 137, United States Department of Interior, Fish and Wildlife Service, Washington, D.C.
5. Persoone, G. and A. Van de Vel. 1987. Cost-analysis of 5 current aquatic ecotoxicological tests. Commission of the European Communities, Report EUR 11342EN.
6. Vanhaecke, P., G. Persoone, C. Claus and P. Sorgeloos. 1980. Research on the development of a short-term standard toxicity test with *Artemia* nauplii. In G. Persoone, P. Sorgeloos, O. Roels and E. Jaspers, Eds., *The Brine Shrimp Artemia.* Vol. 1. Morphology, Genetics, Radiology, Toxicology. Universa Press, Wetteren, Belgium, pp. 263–285.
7. Vanhaecke, P. and G. Persoone. 1984. The ARC-test: a standardized short-term routine toxicity test with *Artemia* nauplii. Methodology and Evaluation. In G. Persoone, E. Jaspers and C. Claus, Eds., *Ecotoxicological Testing for the Marine Environment.* Vol. 2. State University Ghent; Inst. Mar. SCI. Res., Bredene, Belgium, pp. 143–157.
8. Vanhaecke, P., G. Persoone, C. Claus and P. Sorgeloos. 1981. Proposal for a short-term toxicity test with *Artemia* nauplii. *Ecotoxicol. Environ. Safety.* 5:382–387.
9. Snell, T.W. and G. Persoone. 1989a. Acute toxicity bioassay using rotifers. I. A test for brackish and marine environments with *Brachionus plicatilis. Aquatic Toxicol.* 14:65–80.
10. Snell, T.W. and G. Persoone. 1989b. Acute toxicity bioassay using rotifers. II. A freshwater test with *Brachionus rubens. Aquatic Toxicol.*14:81–92.

11. Snell, T.W., B.D. Moffat, C. Janssen and G. Persoone. 1991. Acute Toxicity bioassays using rotifers. IV. Effects of cyst age, temperature and salinity on the sensitivity of *Brachionus calyciflorus*. *Ecotoxicol. Environ. Safety*. 21:308–317.
12. Brendonck, L. 1990. Redescription of fairy shrimp *Streptocephalus proboscideus* (Frauenfeld, 1873) (Crustacea, Anostraca). Biologie (1989). 59:49–57.
13. Brendonck, L. and G. Persoone. 1991. Biological/ecological characteristics of large freshwater branchiopods from endorheic regions and consequences for their use in cyst-based toxicity tests. This volume.
14. Brendonck, L., G. Uyttersprot and G. Persoone. 1990. A culture system for fairy shrimps (Crustacea, Anostraca). *Aquacultural Engineering*. 9:267–283.
15. Freeman, W. 1953. A standardized method for determining toxicity of pure compounds to fish. *Sewage and Industrial Wastes*. 25:845–848.
16. European Economic Commission. 1984. C.2. Acute Toxicity for *Daphnia*. Commission Directive of 25 April 1984 adapting to technical progress for the sixth time Council Directive 67/548/EEC on the approximation of the laws, regulations and administrative provisions relating to classification, packaging and labelling of dangerous substances. Commission 84/449/EEC, Brussels.
17. U.S. Environmental Protection Agency. 1985. Methods for measuring the acute toxicity of effluents to freshwater and marine organisms. EPA-600/4–85–013. Environ. Research Laboratory, Duluth, MN.
18. American Public Health Association, American Water Works Association and Water Pollution Control Federation. 18985. *Standard Methods for the Examination of Water and Wastewater,* 15th ed. American Public Health Association, Washington, DC.
19. Stephan, C.E. 1977. Methods for calculating an LC_{50}. In F.L. Mayer and J.L. Hamelink, Eds., *Aquatic Toxicology and Hazard Evaluation*, American Society for Testing and Materials, Philadelphia, pp. 65–84.
20. Hamilton, M.A., R.C. Russo and R.V. Thurston. 1977. Trimmed Spearman-Karber method for estimating median lethal concentrations in toxicity bioassays. *Environ. Sci. Technol.* 11:714–719; Correction 12:417 (1978).
21. Mitchell, S.A. 1990. Factors affecting the hatching of *Streptocephalus macrourus* Daday (Crustacea, Eubranchiopoda) eggs. *Hydrobiologia*. 194:13–22.
22. Lewis, P.H. and C.I. Weber. 1985. A study of the reability of *Daphnia* acute toxicity tests. In R.D. Cardwell, R. Purdy and R.C. Bahner, Eds., *Aquatic Toxicology and Hazard Assessment*: Seventh Symposium, American Society for Testing and Materials, Philadelphia, pp. 78–86.

23. Florida Department of Environmental Regulation. 1986. Standard Reference Toxicant Bioassays. Second Quarter—April Through June 1986, Florida.
24. Mount, D.I. and T.J. Norberg. 1984. A seven-day life-cycle cladoceran toxicity test. *Environ. Toxicol. Chem.* 3:425–434.
25. Lussier, S.M., J.H. Gentile and J. Walker. 1985. Acute and chronic effects of heavy metals and cyanide on *Mysidopsis bahia* (Crustacea: Mysidacea) *Aquat. Toxicol.* 7:25–35.
26. Qureshi, A.A., R.N. Coleman and J.H. Paran. 1983. Evaluation and refinement of the Microtox test for use in toxicity screening. In B.J. Dutka and D. Liu, Eds., *Toxicity Screening Procedures Using Bacterial Systems.* Vol. 1. Toxicology Series. Marcel Dekker, Inc., New York, pp. 89–118.
27. Microtox™ Application Notes. 1981. Beckman Instruments, Inc., Carlsbad, CA.

CHAPTER 4

Reducing Variability in *Daphnia* Toxicity Tests — A Case for Further Standardization

Mairead C. Bradley, Caroline Naylor, Peter Calow, Donald J. Baird, Ian Barber, and Amadeu M. V. M. Soares

ABSTRACT

Problems associated with reproducibility of *Daphnia* toxicity tests are examined by considering the bioassay as a multicomponent system. The major components of this system are considered separately with the aim of standardization. This is discussed in light of studies of sources of variability in *Daphnia* toxicity tests carried out by the authors and reported in the literature.

0-87371-845-3/93/$0.00 + $0.50

KEY WORDS

Daphnia, bioassay, life-history variation, laboratory culture, maternal environment.

INTRODUCTION

Reproducibility is a key criterion for any laboratory toxicity test but one that is often imperfectly met. This was exemplified by the 1986 'ring test' carried out by the European Commission (EC) to assess interlaboratory consistency in the performance of the 21 d chronic test using *Daphnia magna* Straus.[1,2] Out of the 37 participating laboratories, only 22 met the basic validity criteria specified in the protocol, and the final results showed the maximum possible variation in response.

One way to address this problem is to separate out and examine the different components of variability in bioassay response corresponding to genotype and environment. This approach, adapted from standard methods in quantitative genetics, has been employed previously.[1] Here, we shall investigate the bioassay as a system made up of interacting parts.[3] Our aim is to define a system that can easily be transferred between laboratories.

Figure 1 illustrates the system and how the various components interact. Here, we shall consider only the organism and the culture system, leaving aside the design of the bioassay itself, and the specific problems posed by different classes of toxic compounds. We first concentrate on sources of variability deriving from the organism itself — genotypic effects — and then consider environmental effects, which are dependent on the design and management of the culture system.

THE TEST ORGANISM

Test guidelines (e.g., References 4, 5, and 6) generally designate species for use in laboratory toxicity tests, in this case *Daphnia magna*, but no consideration is made concerning the source of origin nor the genetic variation within the laboratory population. Since *D. magna*'s parthenogenetic (clonal) mode of reproduction is widely quoted as one

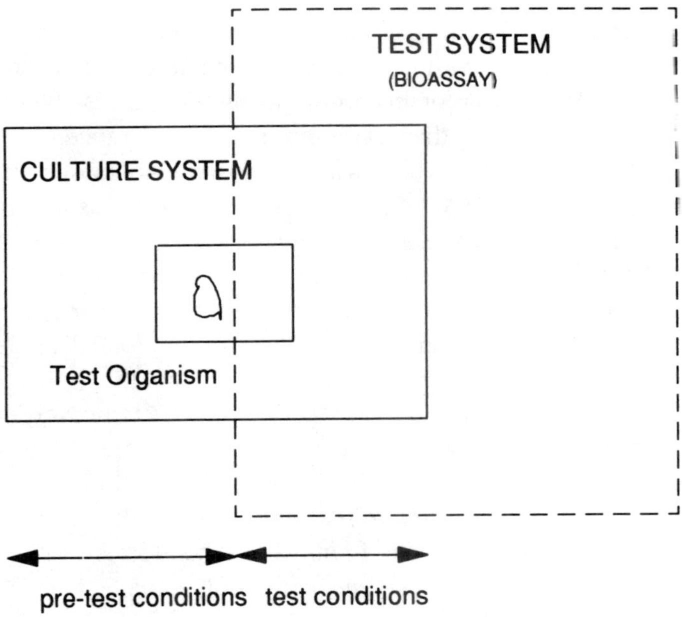

FIGURE 1. For the purposes of standardization, the *Daphnia* bioassay can be considered to be made up of three interlocking systems: (1) the test organism, (2) the culture system, and (3) the system. It can be also be divided into pre-test and test conditions; the test organism being common to both of these divisions.

of its advantages as a test organism, such specification should present no difficulties. The two key questions are: how genetically variable are laboratory stocks, and how relevant is this variability to performing laboratory toxicity tests?

To address the first question, we analyzed genetic variability in 27 of the laboratories involved in the 1986 EC *D. magna* ring test, and demonstrated that all but one of the laboratories cultured single-clone populations (the exception culturing a mixture of two clones).[7]

To address the second question, we compared the acute responses of a selection of these clones to a range of toxicants. This revealed significant differences in the 48-h EC_{50} values; e.g., for cadmium, 0.6 to 140 ppb.[7,8] Although the range in response obtained varied with each toxicant tested, these were often greater than an order of magnitude and therefore of importance for toxicity testing. However, when the same group of clones was compared in terms of the chronic response to a smaller group of compounds,[9] again revealing significant differences in genotype response (in terms of no-effect concentrations), the range in response was always within an order of magnitude, indicating that perhaps interclonal variation in chronic stress tolerance is of less importance for toxicity testing than variation in acute tolerance.

The performance of these clones under constant laboratory conditions (see legend to Figure 2), revealed significant differences in a range of basic life-history traits.[10] As an example, variation in four traits is shown in Figure 2: here, significant interclonal variation was found in time to first reproduction, juvenile growth rate, and capacity for increase. The comparisons were made at two different food densities: 0.33 mg $C.L^{-1}$ and 1.67 mg $C.L^{-1}$; it can also be seen that this environmental factor had a significant influence on the life-history. However, it should be noted that although these differences between clones were statistically significant, they were relatively small in magnitude.

In responses to a questionnaire sent out as part of the interlaboratory genetic typing exercise, it was found that laboratories routinely performing bioassays with *D. magna* commonly obtained their source material from local pond populations. In establishing a laboratory culture, the source material could be considered as a colonizing population passing through a selection 'bottleneck' as it entered in the laboratory environment.[11] Such laboratory populations have been maintained under relatively constant environmental conditions in terms of temperature, light, water quality, and food ration for, in many cases, more than 50 generations.[7] In fact, the laboratory environment provides a selection regime which favors early reproduction (since cultures are periodically reset) and maximal offspring production (and hence also reducing egg size,[12] since food resources are generally provided at *ad libitum* levels). Sexual reproduction would seem to be rare in laboratory populations. This is not surprising, given that genetic typing studies[7,13] have revealed that most laboratory cultures consist of single-clone populations in which ephippial production is rare, and that even if produced, ephippia will be unable to hatch in most cultures due to the periodic resetting.

In the laboratory, genotypes will be selected for production (toxicity tests require large numbers of offspring), shortened developmental time, and robustness (simply by their ability to survive in the laboratory). Since laboratories generally culture single clones, chronic test results based on such material would be generally comparable between laboratories, since measures of chronic toxicity are normally relative measures based on the performance of control animals. For genetically based life-history differences to have a significant impact on test results,

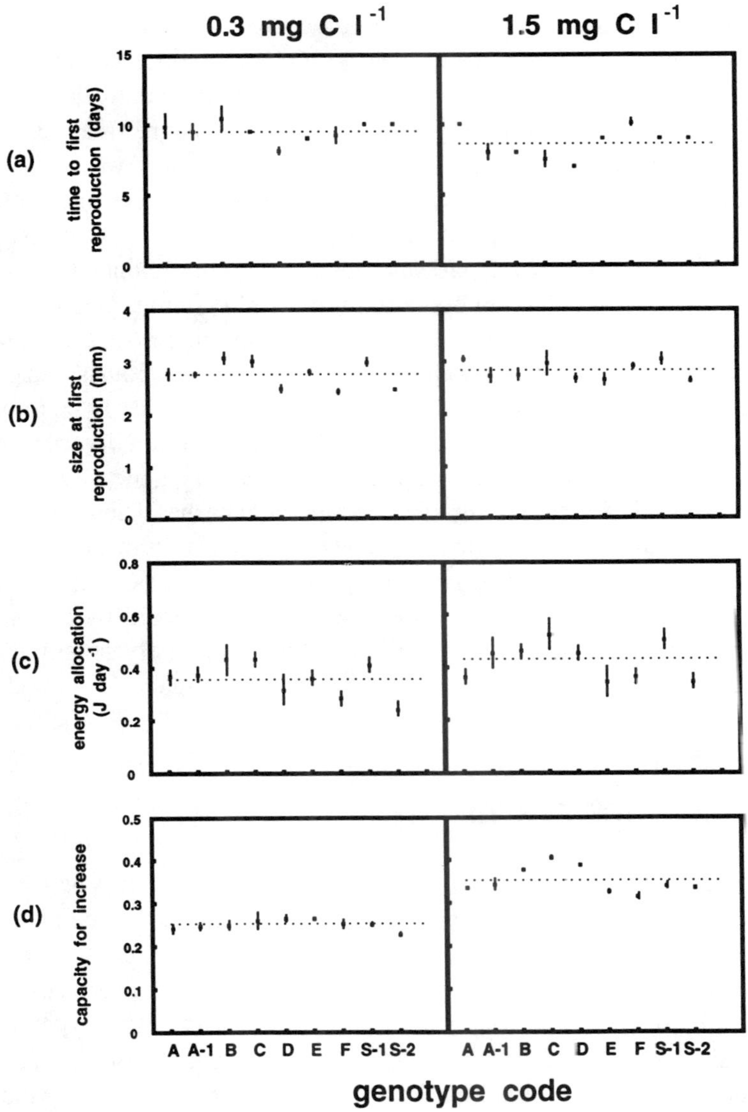

FIGURE 2. Interclonal variation in *D. magna* in four life-history traits at each of two ration levels (0.33 mgC.L^{-1} and 1.67 mgC.L^{-1}). All data are means ±s.d.[10]: (a) time to first reproduction (=clutch release) in days; (b) size at first reproduction (=clutch release) based on body length measurements in mm from the base of the tail spine to the top of the head; (c) energy allocation to juvenile growth (during instars 1–4), based on dry mass/joule-equivalent conversions given in Reference 8; (d) capacity for increase (calculation given in Reference 8). The overall mean response (pooling data from all genotypes) for each trait is indicated in each graph by the horizontal line.

resulting in an over- or underestimation of toxicity, it would be necessary to combine different genotypes within the test in a systematically biased mode (e.g., a low fecundity genotype as the control combined with a high fecundity genotype used in the exposure treatments), a situation which is unlikely to arise. For this reason, it would be less important to specify the genotype of *Daphnia* used in chronic toxicity tests (although if this can be done, so much the better, since it may account for approximately 20% of total variability[9]), than it is to be certain that cultures do not contain mixtures of different genotypes. In turn, it will be very important to control environment (that is, culture conditions) than to worry about genetically based inconsistencies in results between laboratories, since the latter will most likely be minor.[9]

Initially, it was hoped that investigations of interclonal variation in tolerance to acute and chronic stress would identify generally tolerant genotypes. However, no significant concordance has been found between the rank order of sensitivity between clones, both in acute and chronic responses, and also in genotype acute-chronic performance.[7–10] For this reason, a pragmatic approach is being adopted by European laboratories: selection of a single clone that is relatively sensitive and has been maintained in culture for a number of years.

THE CULTURE SYSTEM

The primary aim of any culture system for *D. magna* to be used in toxicity tests is the *reliable* production of large numbers of neonates of a *consistent quality*. A wide range of culture systems is currently employed, meeting these criteria with varying degrees of success.

In order to consider *D. magna* culturing in relation to toxicity testing, several points should be emphasized. First, culture conditions can be divided into pre-test 'maternal' conditions (stock cultures, producing neonates to be used in tests) and the conditions in the test itself[1] — see Figure 1. Currently, guidelines specify (to a limited extent) only the latter, and consequently laboratories could be considered to run two different sets of culture conditions: stock cultures are generally run for convenience, whereas test cultures are run to fit in with legislative demands. Since the aims of the culture conditions both before and during the tests are essentially the same (that is, the successful maintenance of healthy, reproducing daphnids), it is surely logical to consider the possibility of adopting a single culture system.

The significant influence of the maternal environment is well illustrated by considering the effect of a daphnid's ration level on the performance of her offspring in acute toxicity tests. We have shown that daphnids held at low ration produce fewer neonates, but that these neonates are larger (in terms of body length and mass) and are relatively more tolerant to acute stress from a number of compounds[7,14] (see also Enserink[15]). This is illustrated in Figure 3 for sodium bromide acute stress.

There are several difficulties associated with the proposed standardization of *D. magna* culture systems. The first is perhaps the least problematic; laboratories often maintain population or 'bulk' pre-test cultures, and may carry out tests with individual animals. The reasons underlying this are basically sound: individual cultures allow more

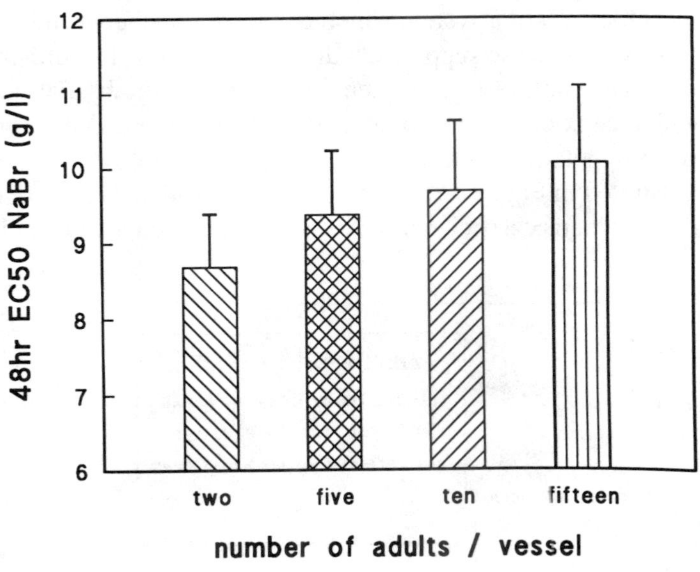

FIGURE 3. The effect of maternal ration on acute response of neonates to sodium bromide. Maternal ration was manipulated indirectly by varying the stocking density in bulk cultures. Groups of 2, 5, 10, or 15 daphnids were maintained in 800 ml ASTM (hard) water in 1.0 l glass vessels, with the addition of an organic additive[3] and kept at 20°C on a 14:10 light:dark cycle. Daphnids were fed with a daily ration of *Chlorella vulgaris* at 0.2 mg C *per vessel*, per day. Effective ration/daphnid therefore decreased with increasing stocking density. Neonates were removed at less than 24 h old and standard 48-h acute tests carried out using 3rd, 4th, and 5th brood neonates. Representative EC_{50} data for acute response to NaBr[14] are shown.

detailed information to be gained from tests, whereas population cultures are less labor-intensive, and hence less costly. We see no difficulty in maintaining this two-tier system, assuming major system parameters are held constant: that is media composition, additives, and food ration (in terms of carbon density of the medium). We shall discuss these further.

CULTURE ADDITIVES

A variety of substances are often added to cultures to improve performance; e.g., seaweed extracts, soil extracts, yeasts, activated sludge, vitamins, dried fish foods (Tetramin like). The presence of these essentially undefined components in the culture system is a major constraint to the development of a system which is transferable between laboratories. Additives may also supplement dietary carbon levels. Furthermore, they may cause unforeseen problems in toxicity tests through interactions with the toxicity of the compound being tested, e.g., chelators will affect heavy metal bioavailability.

As illustrated in Figure 4, additives can be considered to make up for media and/or dietary deficiencies. Although improved understanding of

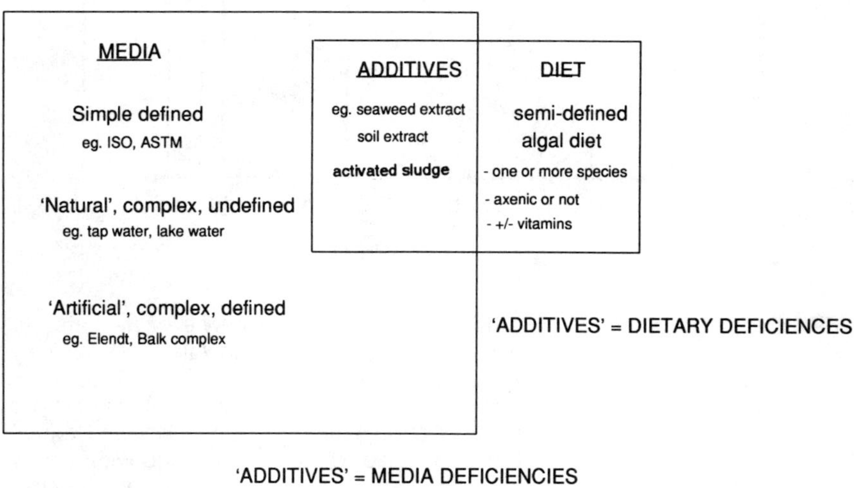

FIGURE 4. Additives, widely used in daphnid culture systems, can be considered as either making for deficiencies of diet or media.

the algal component of the diet may allow additives to be dispensed with, progress in this area to date has instead focused on improving media quality. Removing additives through improvement in media quality has the advantage that the dietary component of the systems should then be relatively simple to standardize. Since *D. magna* will thrive on a variety of micro-algal species,[16] designation of one or more species (previously assessed for dietary suitability) and a standard culture media should be straightforward.

CULTURE MEDIA

The poor reproduction and survival of cultures of *D. magna* in completely defined media (that is, artificial pond waters), have resulted in many laboratories culturing their animals and conducting tests in natural waters (that is, a variety of surface waters and groundwaters). This too, however, can present problems for test standardization: the composition of these waters is often unspecified, and could conceivably vary over time (although it should be added that many are carefully monitored). Trace contaminants in tap waters, for example, could acclimate daphnids to stressors.[17]

The successful development of a completely defined synthetic culture medium for *D. magna*, suitable also for testing purposes, is the way forward. Many attempts have been made to produce defined *Daphnia* media (e.g., References 6, 18, and 19), although the majority of these are simple media composed of a few salts, all appearing to be deficient in some factors. The complex defined medium of Keating[20] has the major drawback of containing an organic buffering agent, glycylglycine, which provides an excellent substrate for bacterial growth. Two other complex defined media have recently been the subject of investigation: 'Balk complex' medium,[21] based on ISO medium, and 'Elendt M4' medium,[22] developed from Keating's medium (recipes for both are given in Table 1).

Figure 5 presents data comparing the performance of 'Elendt M4' medium with the medium used routinely in some of our laboratories (ASTM + seaweed extract), ASTM only, 'Balk simple' and 'Balk complex'. 'Balk simple' is another synthetic medium composed of ISO water with trace amounts of selenium and bromine added.[21] Illustrated is mean fecundity per vessel for daphnids acclimated to each medium

Table 1.
Key Components of Balk Complex and Elendt's M4 Media

Trace element	Balk complex	Elendt
Br, Sr, I, Se, Rb, V, Li	Yes	Yes
F, Ni, Al, Cr, Pb	Yes	No
Si, Fe, Mn, Mo, Cu, Zn, Co, B	No	Yes

Notes: (1) Both the complex medium suggested by Balk and Elendt's M4 medium use the ISO 6341 medium as base. (2) In addition to ISO medium, Balk complex contains compounds prepared as 2 stock solutions. (3) Elendt medium contains 22 compounds, all prepared separately and includes vitamins and EDTA.

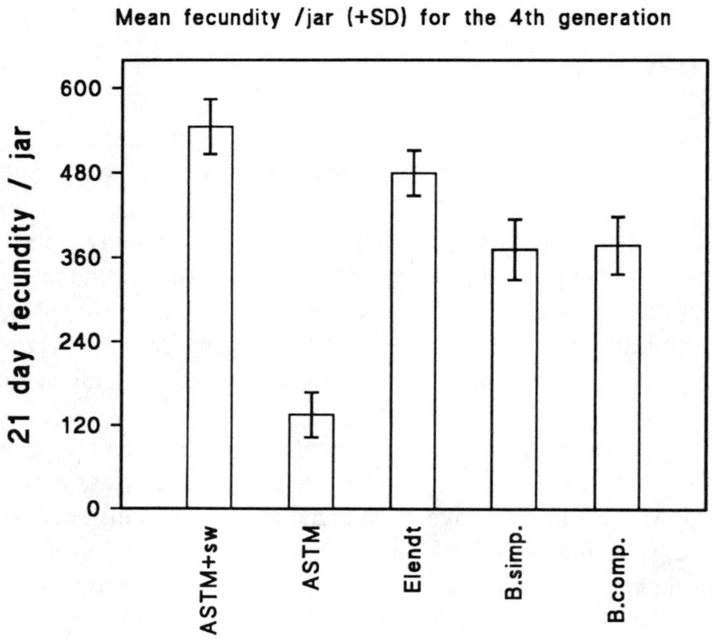

FIGURE 5. Long term performance of a single clone of *D. magna* in bulk culture in five different media. Data are mean cumulative fecundity (±s.d.) over 21 d *per vessel*. Vessels were initially set up with 5 individuals in 800 ml of medium and given a food density of 1.25 mgC.L^{-1}, changing on alternate days. Ration *per vessel* was maintained at the same level over the experiment. Cultures were maintained at 20°C on a 14:10 light:dark cycle. Data are for 4th generation cultures, each generation consisting of a 21 d period after which cultures were re-set from neonates. B: Balk; SW: Seaweed additive — see text.

and cultured under identical conditions (see figure legend) over 21 d, the duration of a standard *D. magna* chronic test. This shows that the performance of daphnids cultured in Elendt M4 medium was not significantly different from those cultured in ASTM + seaweed extract. Performances using the two Balk media were not significantly different from each other, but poorer than in all other media with the exception of ASTM alone, which gave the poorest performance of all.

Since survival was good in all but the ASTM-only medium, these values can be divided through by the number of daphnids (that is, 5) to compare performance with the EC validity criteria for fecundity in the reproduction test, that is, 60 neonates over the duration of the test. On this basis, both ASTM + seaweed extract and Elendt M4 easily satisfied the fecundity validity criterion. Both the Balk media resulted in average fecundities exceeding, but nonetheless close to the threshold of the validity criterion. Mortality in the ASTM-only treatment resulted in this medium failing the survival validity criterion, although it was clear that reproduction was extremely poor as well.

Further work is required to develop Elendt M4 medium for routine use in *D. magna* culturing, particularly to reduce the level of EDTA. In addition, it should also be noted that artificial media are only as good as their matrix water, and contamination of nanopure waters with resin residues from ion exchangers can be a problem. However, despite these problems, studies carried out so far on Elendt M4 medium are promising.

DIET

The ideal method of standardizing the diet of *D. magna* would be to develop a completely artificial diet. Since, at present, little is known concerning the dietary requirements of microcrustaceans, chemically defined artificial diets are not available.[16] In their absence, it is important to standardize the variety of natural diets currently used. Important here would be standardization of the amounts of energy and mass available to each daphnid. Although there are other dietary factors which will affect performance, such as algal digestibility or vessel size effects,[4,16] it is clear that food availability is potentially a major source of variation in toxicity tests.[4]

Recent results of a European interlaboratory standardization exercise has suggested that if ration is specified in terms of carbon mass, then factors such as algal species or culture technique do not significantly affect daphnid performance. A ration of between 0.1 and 0.4 mg C.daphnid^{-1}.day^{-1} has been proposed, which (it is claimed) will easily allow sufficient reproduction to satisfy the validity criterion, while preventing the health problems associated with overfeeding.

In practice, it may be diffucult to measure carbon on an ongoing basis, and here it is worth noting that measures of cell number or cell absorbance can be good predictors of carbon content of an algal culture. One major caveat is that these relationships will be culture-system dependent, and not easily transferable: laboratories should use their own nomographs. In addition, potential temporal variation in cell size should be monitored in cultures where possible.

As a final cautionary note, there are also problems associated with the use of mg C.daphnid^{-1}.day^{-1} to specify food level: daphnids vary their feeding rate in response to food density, not to absolute amounts of food. So, for this reason culture volumes must be held constant in order to maintain similar feeding rates between experiments and laboratories, and be sufficiently large to prevent food depletion effects. By the same token, if animals die in cultures, the volumes must be adjusted to maintain a constant mass density for the remaining animals.

CONCLUSIONS

Standardization of both the *D. magna* culture and test systems is now largely accepted and indeed currently underway in Europe. Although biological studies and interlaboratory tests involving *D. magna* will continue, the progress which has recently been made suggests that such standardization will ultimately lead to a simpler, more cost-effective and reliable test system which is transferable between laboratories.

ACKNOWLEDGMENTS

Portions of this work were supported by grants from EC DG XI, EC DG XII (grant B/86000160), the Natural Environment Research Council (grant TSF/86/AeT/4), NATO (grant 3/A/87/PO), Unilever Research Ltd., and Shell Research Ltd. Thanks also go to the technicians of the *Daphnia* Unit at Sheffield University, past and present.

REFERENCES

1. Baird, D.J., I. Barber, M.C. Bradley, A.M.V.M. Soares and P. Calow. 1989. The *Daphnia* bioassay: a critique. *Hydrobiologia* 188/189: 403–406.
2. Cabridenc, R. 1986. Exercice d'intercalibration concernant une méthode de determination de l'écotoxicité a moyen terme des substances chimiques vis-a-vis des daphnies. Unpublished EC Report. Contract W/63/476 (214).
3. Baird, D.J., A.M.V.M. Soares, A. Girling, I. Barber, M.C. Bradley and P. Calow. 1988. The long-term maintenance of *Daphnia magna*, Straus, for use in ecotoxicity tests: problems and prospects. In H. Lokke and F. Bro-Rasmussen, Eds, *Proceedings of the First European Conference on Ecotoxicology*. Lyngby, Denmark, pp. 144–148.
4. EC. 1979. Amending for the sixth time directive 67/548/EEC on the classification, packaging and labelling of dangerous substances. Council Directive 79/831/EEC. OJL259/10. Community of the European Community, Brussels.
5. OECD. 1981. *Daphnia* sp. 14 day reproduction tests (including an acute immobilisation test) (Guidelines for the testing of chemicals no.202.). OECD, Paris.
6. ISO. 1982. Water quality — determination of the inhibition of the mortality of *Daphnia magna* Straus (Cladocera:Crustacea) (Technical report 6241). International Standards Organisation.
7. Baird, D.J., I. Barber, M.C. Bradley, A.M.V.M. Soares and P. Calow. 1991. A comparative study of genotype sensitivity to acute toxic stress using clones of *Daphnia magna* Straus. *Ecotoxicol. Environ. Saf.* 21: 257–265.
8. Baird, D.J., I. Barber and P. Calow. 1990. Clonal variation in general responses of *Daphnia magna* Straus to toxic stress. I. Chronic life-history effects. *Funct. Ecol.* 4:399–408.
9. Soares, A.M.V.M., D.J. Baird and P. Calow. 1992. Interclonal variation in the performance of *Daphnia magna* Straus in chronic bioassays. *Environ. Toxicol. Chem.* 11:1477–1483.

10. Soares, A.M.V.M. 1989. Clonal Variation in Life-History Traits in *Daphnia magna* Straus (Crustacea, Cladocera). Implications for Ecotoxicology. Ph.D. dissertation, University of Sheffield, Sheffield, U.K.
11. Baird, D.J. 1992. Predicting population response to pollutants: in praise of clones. *Funct. Ecol.* 6:616–619.
12. Glazier, D.S. 1992. Effects of food, genotype and maternal size and age on offspring investment in *Daphnia magna. Ecology* 73:910–926.
13. Snyder, T.P., K.M. Switzer and R.E. Keen. 1991. Allozymic variation in toxicity-testing strains of *Ceriodaphnia dubia* and in natural populations of *Ceriodaphnia. Environ. Toxicol. Chem.* 10:1045–1049.
14. Naylor, C., E.J. Cox, M.C. Bradley and P. Calow. 1992. Effect of differing maternal food ration on susceptibility of *Daphnia magna* Straus neonates to toxic substances. *Aquat. Toxicol.* 24:63–74.
15. Enserink, L., W. Luttmer and H. Maas-Diepeveen. 1990. Reproductive strategy of *Daphnia magna* affects the sensitivity of progeny in acute toxicity tests. *Aquat. Toxicol.* 17:15–25.
16. Vijverberg, J. 1989. Culture techniques for studies on the growth, development and reproduction of copepods and cladocerans under laboratory and *in-situ* conditions: a review. *Freshwater Biol.* 21:317–373.
17. Stuhlbacher, A., M.C. Bradley, C. Naylor and P. Calow. 1992. Induction of cadmium tolerance in two clones of *Daphnia magna* Straus. *Comp. Biochem. Physiol.* 101C:571–577.
18. ASTM. 1988. Standard guide for conducting renewal life-cycle toxicity tests with *Daphnia magna.* E 1193–87 Annual Book of ASTM Standards, Vol. 11.04.
19. DIN. 1982. Deutsches Institut für Normung. Desutsche Eineitsverfahren Zur Wasser-, Abwasser- und Schlammuntersuchung. Testverfahren mit Wassroganismen (Gruppe L). Bestimmung der Wirkung von Wasserinhaltsstoffen auf Kleinkrebse (Daphnienkurzzeittest) (L11). DIN 38412 Teil 11.
20. Keating, K.I. 1985. A system of defined (*sensu stricto*) media for *Daphnia* (Cladocera) culture. *Archi. Hydrobiol.* 3:321–327.
21. Span, A.S.W., Balk F. and Blok J. 1988. Optimization of the chronic toxicity test with *Daphnia magna*; Part III composition of a synthetic medium. Akzo Research, Subject Report No. R 88/18/0842B, Corporate Research Department, Arnhem.
22. Elendt, B.P. and W.R. Bias. 1990. Trace nutrient deficiency in *Daphnia magna* cultured in standard medium for toxicity testing. Effects of the optimization of culture conditions on life-history parameters of *D. magna. Water Res.* 24(9):1157–1167.

CHAPTER 5

Evaluation of the New "Enzymatic Inhibition" Criterion for Rapid Toxicity Testing with *Daphnia magna*

C.R. Janssen, E.Q. Espiritu, and G. Persoone

ABSTRACT

The recently developed 1-h *Daphnia magna* bioassay was further assessed. This method, which is based on the *in vivo* observation of the inhibition of an enzymatic process, was used to assess the toxicity of nine pure chemicals and seven complex effluents. The results obtained with the new method (1-h EC_{50}s) were compared to those of the conventional acute *D. magna* toxicity test (24-h and 48-h EC_{50}s). For the nine chemical compounds, the 1-h EC_{50}s correlated well with the 24-h and 48-h EC_{50}s; however, no significant correlation was found between both test criteria for the complex effluents. The utility and limitations of the 1-h assay are discussed in light of the results.

0-87371-845-3/93/$0.00 + $0.50

KEY WORDS

Aquatic toxicology, *Daphnia magna*, test method, toxicity screening, enzymatic criterion.

INTRODUCTION

The most well-developed and widely used test methods for toxicity assessement of chemicals and effluents are acute bioassays. However, for routine applications such as wastewater monitoring and in-house screening of newly developed chemical compounds, these conventional tests are often too time consuming and costly. Indeed, the two most widely used acute tests, i.e., the fish and invertebrate toxicity test, cost in excess of $700 U.S. and require 48 to 96 h to complete.[1] The need for faster and less expensive tests for routine determinations of acute toxicity has been expressed by various authors[2–4] and has, in the last few years, led to the development of alternative (faster) toxicity test methods such as short-term bacterial and *in vitro* screening tests. Various bacterial bioassays have been developed based on the measurement of different criteria related to the biological status of the microbial systems, e.g., enzymatic activity, growth inhibition, reproduction ratio, oxygen demand, metabolic light, and heat release.[5,6] *In vitro* screening bioassays, on the other hand, use cellular or sub-cellular preparations to examine the potential toxicity of chemicals and are usually based on the measurement of membrane-related processes, enzymatic activity and electron transport mechanisms.[1]

The major advantage associated with both types of toxicity screening tests is their rapidity; i.e., test results are available within hours (minutes in some cases). However, several factors, such as the availability of specialized equipment and/or the biological material, hamper the routine use of these screening bioassays. Moreover the utility and acceptance of these "alternative" methods depends to a large extent on their predictive potential; i.e., their level of correlation with indices of acute toxicity obtained with conventional (eukaryotic whole-organism) testing methods.

In this context, the recently developed 1-h *Daphnia magna* bioassay appears to be a promising method for rapid routine toxicity evaluations.[7] This methodology requires no sophisticated instrumentation or specific technical skills and is completed in less than 1 1/2 h. Moreover, this screening bioassay uses the neonates of *D. magna* — the most widely used test organism in acute toxicity evaluations — as test organisms. In this chapter, the utility of this method is further evaluated. Results obtained with the 1-h "enzymatic inhibition" test are compared to those with the conventional 24- and 48-h test methods, for 9 chemicals and 7 complex effluents.

BACKGROUND AND PRINCIPLE OF THE METHOD

The new toxicity screening method is based on the *in vivo* observation of the inhibition of (a) certain enzymatic process(es) in the daphnids. It is derived from well-established techniques for measuring the enzymatic activity of bacterial populations.[6,8,9] These techniques use chromogenic or fluorogenic substrates that, after cleavage by the corresponding enzyme, release the originally bound dye. This can then be detected with spectrophotometer or spectrofluorometer. The more substrate that is cleaved the more dye (colored or fluorescent) is released.

Here, the fluorogenic substrate 4-methylumbelliferyl-β-D-galactoside (MFU-galactoside) is used as a sublethal stress indicator of *in vivo* enzymatic inhibition in the freshwater crustacean *Daphnia magna*. This MFU-substrate, which is added to the test solution (and ingested by the daphnids) after an initial exposure period of 1 h to the toxicant, yields upon enzymatic hydrolysis, 4-methylumbelliferone which appears as a strongly fluorescent dye throughout the daphnid's body. The cleavage of the MFU-substrate is the toxicity response criterion and the inhibition of the enzymatic hydrolysis is dose-dependent.

EXPERIMENTAL DESIGN

The experimental protocol, which is represented schematically in Figure 1, basically consists of exposing juvenile *Daphnia magna* to a series of toxicant dilutions (without food) for 1 h, after which the substrate is added. Fifteen minutes later, the number of fluorescent test animals is counted using a U.V. light, and the 1-h EC_{50} is calculated.

Each toxicity test consists of a control and five toxicant treatments. Six 10-ml glass test tubes with five *D. magna* neonates (<24 h old) are used per treatment. The tubes are incubated in darkness for 1 h at 20°C,

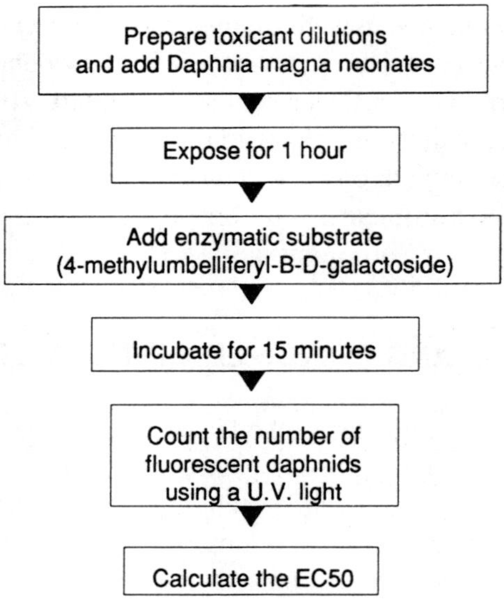

FIGURE 1. Schematic representation of the 1-h EC50 *Daphnia magna* toxicity test procedure.

after which 0.1 ml of a 5000 mg/l MFU-substrate stock solution is added to each test tube. After 15 min of substrate incubation, the number of fluorescent animals are counted using a long wave U.V. light (385nm). A test is considered valid if 90% of the control animals exhibit a positive response (fluorescence).

The MFU-substrate stock can be prepared by adding 50 mg of 4-methylumbelliferyl-β-D-galactoside powder to 10 ml of deionized water; this solution is then shaken vigorously for 3 min.

Concurrent with the 1-h fluorescence tests, conventional 24- and 48-h acute toxicity tests with *D. magna* have been carried out using the same test design as described above and following the OECD Guidelines.[10]

Details of the *D. magna* culturing conditions are reported in Reference 7.

CHEMICAL COMPOUNDS — EFFLUENTS

Nine chemical compounds have been tested: urea, acetic acid, trichloroacetic acid, n-propanol, chloral hydrate, 3,4-dichloroaniline (3,4-DCA), lindane, EDTA, and the linear alkyl sulfonate:LAS $(C_{11,18})$.

Reagent grade chemicals were dissolved in reconstituted moderately hard artificial water, prescribed by the U.S. EPA.[11] Acetone at a concentration of 0.25 ml/l was used as a solvent for lindane and 3,4-DCA. For these chemicals, solvent controls (0.25 mg/l) were tested and no adverse effects of the acetone were observed. After preliminary range-finding testing, final test solutions spanned a one log concentration gradient. All concentrations reported are nominal.

The seven complex effluents tested originated from a pharmaceutical company and were taken after the secondary treatment system, at regular intervals. Each sample was divided into different aliquots and stored in a deep-freezer. The enzymatic inhibition assay and the conventional test were always carried out concurrently on the same aliquot.

All EC_{50}s were calculated using probit analysis and are expressed in mg/l for the pure compounds and in percentage for the effluents.

RESULTS AND DISCUSSION

Table 1 gives the 1-h EC_{50}s (for the enzymatic inhibition criterion) and 24- and 48-h EC_{50}s (immobility criterion) for the 9 chemicals tested. The 48-h EC_{50}s are lower than the 1-h EC_{50}s for 7 of the 9 compounds, while the 24-h results are lower in 4 of the 9 cases. All EC_{50}s obtained with the new test method are, however, of the same order of magnitude as those obtained with the conventional bioassays, except for the LAS results and the 48-h EC_{50} for 3,4-DCA. The regression analysis between the results of both test methods gave the following equations: $\log(1h\ EC_{50}) = -0.092 + 1.068 \log(24h\ EC_{50})$ and

Table 1
Comparison of the *D. magna* 1-h EC_{50} (new criterion) Values with the *D. magna* 24-h and 48-h EC_{50}'s (conventional test) for Nine Chemicals.

Test substance	Number of tests	1-h EC_{50} Mean	1-h EC_{50} S.D.	24-h EC_{50} Mean	24-h EC_{50} S.D.	48-h EC_{50} Mean	48-h EC_{50} S.D.
Urea	1	4070	—	6390	—	3910	—
Acetic acid	2	76	18	71	3.0	65	9.0
Trichloroacetic acid	2	137	15	164	28	146	12
n-Propanol	2	2200	221	4850	91	4620	22
Chloral hydrate	2	170	22	246	35	112	12
3,4-DCA	2	1.56	0.28	1.22	0.12	0.42	0.03
Lindane	2	1.46	0.21	1.99	0.21	1.99	0.15
EDTA	1	129	—	122	—	113	—
LAS ($C_{11,18}$)	1	117	—	49	—	33	—

Note: All results are expressed in mg/l.

$\log(1\text{h } EC_{50}) = -0.298 + 1.098 \log(48\text{h } EC_{50})$, respectively. The r^2s for these equations are 0.98 and 0.96, respectively, indicating a good correlation between both effect criteria. In their initial paper on this new test methodology, Janssen and Persoone,[7] using the same test design, reported a similar good correlation for 11 pure compounds, with r^2s of 0.93 and 0.92 for the 1-h vs. the 24-h EC_{50} and the 1-h vs. the 48-h EC_{50} comparisons, respectively.

The regression analysis between the 1-h EC_{50}s and 48-h EC_{50}s for the combined data sets (data of the present study and those of Janssen and Persoone,[7]) is illustrated in Figure 2. The r^2 for this regression is 0.94 ($\log(1\text{h } EC_{50}) = 0.152 + 0.956 \log(48\text{h } EC_{50})$); ($p<0.05$). Similarly, an r^2 of 0.95 was found for the 1-h vs. 24-h comparison ($\log(1\text{h } EC_{50}) = -0.143 + 1.014 \log(24\text{h } EC_{50})$).

The results for the 7 complex effluents tested with the new and the conventional test methods are presented in Table 2. The 1-h EC_{50}s are lower than the 24-h EC_{50}s in all but one case, and lower than the 48-h EC_{50}s in 5 of the 7 effluents tested. The enzymatic inhibition assay indicated toxicity in effluents C and G, while the 24-h conventional toxicity test did not reveal any adverse effects for those effluents. Linear

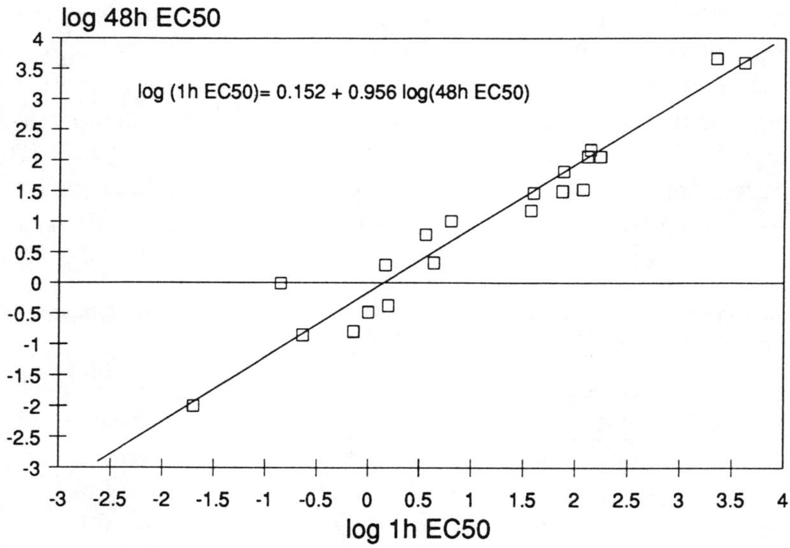

FIGURE 2. Linear regression of the log 1-h EC50s (new criterion) vs. the log 48–h EC50s (conventional test) for 20 chemical compounds (data from this study and Janssen and Persoone, 1991).

Table 2
Comparison of the *D. magna* 1-h EC_{50} (new criterion) Values with the *D. magna* 24-h and 48-h EC_{50}'s (conventional test) for Seven Effluents Originating from a Pharmaceutical Plant.

		1-h EC_{50}		24-h EC_{50}		48-h EC_{50}	
Effluents	**Number of tests**	**Mean**	**S.D.**	**Mean**	**S.D.**	**Mean**	**S.D.**
A	2	19.0	3.8	2.0	0.40	1.2	0.25
B	2	25.6	0.7	33	5.1	15	2.4
C	2	32.2	6.8	—[a]	—	—[a]	—
D	2	29.0	12	47.8[b]	—	42.1[b]	—
E	2	21.2	0.2	61	5.0	59	6.8
F	2	52.1	14	80	5.0	60	7.4
G	2	59.6	14	—[a]		65.3	3.0

Note: All results are expressed in percent effluent.

[a] No mortality observed.
[b] Only one toxicity test performed.

regression analysis on this (limited) data-set showed that neither the 1-h EC_{50} vs. 24-h EC_{50}, nor the 1-h EC_{50} vs. 48-h EC_{50} regressions were significant.

The lack of any correlation found for these effluents between both effect criteria was in sharp contrast to what was observed with the 20 pure chemicals tested so far. Also, the fact that in 6 of the 7 cases the enzymatic inhibition assay was consistently more sensitive than the conventional acute *D. magna* test (which was not the case for pure compounds) requires some attention.

Since the effluents had not been filtered before testing, they still contained high levels of suspended organic material (algae, bacteria, and detritus). It could be that the daphnids fed on this food source during the 1-h exposure period and were fully satiated by the time the MFU-substrate was added. Reduced feeding on the substrate (due to the presence of the suspended organic material in the effluents) would mean less fluorescent animals and thus a lower 1-h EC_{50}. Thus, interference in the MFU-substrate uptake by the presence of natural food particles could possibly (partly) explain the observed results. If this is the case, this new screening bioassay would have to be used with extreme caution when testing wastewaters containing natural food sources for the daphnids, and filtration of the effluents prior to testing would become a prerequisite.

In this context, Janssen and Persoone[7] already suggested that the observed "enzymatic inhibition" criteria might not be the result of biochemical changes in the test organisms, but might be caused by a decrease in the daphnids' feeding rate with increasing toxicant concentrations. The less fluorogenic substrate is ingested by the animal, the less substrate will be hydrolyzed resulting in slightly or non-fluorescent test organisms. This hypothesis has since been tested for some toxicants and preliminary results indicate that although reduced feeding activity is observed with increasing toxicant concentrations, it does not occur at those levels at which the 1-h EC_{50} is found,[12] thus rejecting the hypothesis that the observed "enzymatic inhibition" process is a measure of food-uptake reduction.

Janssen and Persoone[7] reported that the precision of the 1-h bioassay is quite acceptable, with coefficients of variation for the replicate tests (n = 3) ranging from 3 to 32%. The present study revealed the same satisfactory precision for the 1-h toxicity tests (n = 2) with the chemical compounds, with C.V.s ranging from 10 to 24%. The repeatability of the conventional tests was slightly better, with coefficients of variation ranging from 4 to 17% and from 1 to 14% for the 24-h and 48-h EC_{50}s, respectively.

The precision of the 1-h toxicity tests with the effluents was considerably lower than that found for the pure chemicals. Indeed, for six of the seven effluents, the coefficients of variation were higher than 20% (maximum 43%). The repeatability of the conventional effluent toxicity tests, on the other hand, was good with C.V.s ranging from 5 to 20%. Since the high variability between the 1-h effluent tests was not observed in the conventional tests performed on the same effluent aliquots, and since the replicated tests were conducted within a 4-d period, we feel that the observed variability was not due to possible changes in the effluent toxicity (with increasing storage time). The high C.V.s found for the 1-h EC_{50}s could possibly be attributed to differences in the quantity of organic matter present in the different aliquots of the same effluent sample (cf. above).

To assess the utility of the 1-h assay to predict the range of toxicant concentrations in which mortality will occur in conventional *D. magna* tests, the percentage of 24-h and 48-h EC_{50} test results which are situated within a two- and fourfold interval of the 1-h EC_{50}s for the 20 pure chemicals tested to date are plotted in Figure 3 (data from this study and Reference 7).

The 24-h EC_{50}s are within a factor of two of the 1-h EC_{50} in 75% of the cases; 50% of the conventional tests results are higher and 25% are lower than the 1-h EC_{50} results. 60% of the EC_{50}s are situated within this

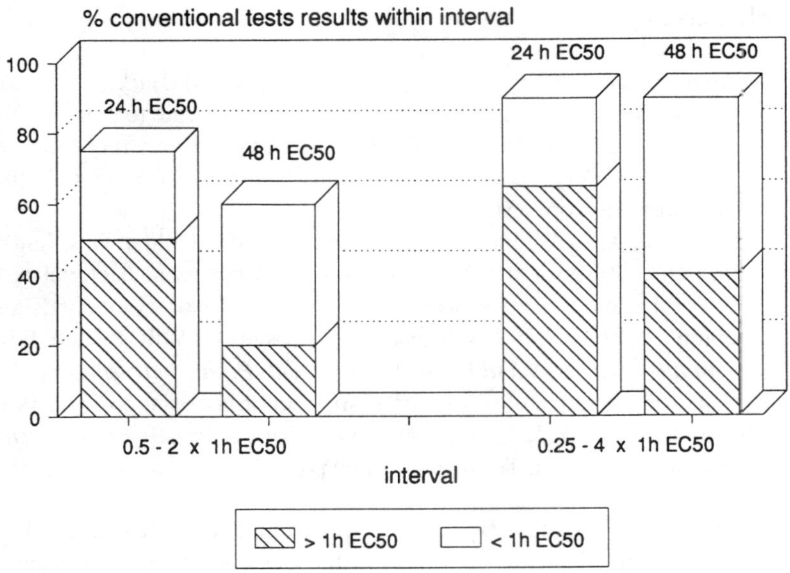

FIGURE 3. Percentage of conventional test results which are situated within a factor of 2 (left) or factor of 4 (right) of the 1-h EC50s for 20 pure chemicals (data from this study and Janssen and Persoone, 1991).

same "factor of two" interval; one-third are higher and two-thirds are lower than the results of the 1-h assay. 90% of the conventional 24-h and 48-h results are within a factor four of the 1-h EC_{50}s. From this, it can be concluded that the 1-h enzymatic inhibition assay does not only correlate well with the conventional *D. magna* acute toxicity results (as shown in Figure 2), but it also has approximately the same sensitivity as the conventional immobility (mortality) test.

This study has shown that the 1-h enzymatic inhibition test could be a valuable tool for the routine screening of chemicals especially when considering the many advantages inherent to this method such as experimental simplicity, short exposure time, and low costs of performance. However, for the application of this methodology for effluent monitoring, we feel that more research is needed to elucidate some of the questions previously raised. Also, further fundamental research is needed to unravel the exact processes that take place during this *in vivo* "enzymatic inhibition" assay with *D. magna*.

REFERENCES

1. Blondin, G.A., L.M. Knobeloch, H.W. Read, and J.M. Harkin. 1989. An *in vitro* submitochondrial bioassay for predicting acute toxicity in fish. In G.W. Suter and M.A. Lewis, Eds., *Aquatic Toxicity and Environmental Fate:* Vol. 11, ASTM STP 1006, American Society for Testing and Materials, pp. 551–563.
2. Qureshi, A.A., K.W. Flood, K.W. Janhurst, S.M. Janhurst, C.S. Inniss, and D.A. Roskosh. 1982. Comparison of luminescent bacterial tests with other bioassays for determining toxicity of pure compounds and complex effluents. In J.G. Pearson, R.B. Foster and W.E. Bishop, Eds., *Aquatic Toxicology and Hazard Assessment: Fifth Conference*, ASTM STP 766, American Society for Testing and Materials, pp. 179–195.
3. Ribo, J.M. and K.L.M. Kaiser. 1987. *Photobacterium phosphoreum* toxicity bioassay. I. Test methods and procedures. *Tox. Assess.* 2:305–323.
4. Mc Feters, G.A., P.J. Bond, S.B. Olson, and Y.T. Tchan. 1983. A comparison of microbial assays for the detection of aquatic toxicants. *Water Res.* 17: 1757–1762.
5. Elnabarawy, M.T., R.R. Robideau, and S.A. Beach. 1988. Comparison of three rapid toxicity test procedures: Microtox, Polytox, and activated sludge respiration inhibition. *Tox. Assess.* 3: 361–370.
6. Obst, U., A. Holzapfel-Pschron and M. Weigand-Rosinus. 1988. Application of enzyme assays for toxicological water testing. *Tox. Assess.* 3: 81–91.
7. Janssen, C.R. and G. Persoone. 1991. Rapid toxicity screening tests for aquatic biota. I. Methodology and experiments with *D. magna*. *Environ. Toxicol. Chem.* (in press).
8. Buikema, A.L. Jr., C.L. Ruther Ford and J. Cairns, Jr.. 1980. Screening sediments for potential toxicity by *in vitro* inhibition. In R.A. and A.Arbor, Eds., *Contaminants and Sediments.* Science Publishers, Inc., Ann Arbor, MI, pp. 463–475.
9. Holzaphel-Pschorn, A., U. Obst and K. Haberer. 1987. Sensitive methods for the determination of microbial activities in water samples using fluorogenic substrates. *Fresenius Z. Anal. Chem.* 321:166–168.
10. OECD. 1984. *Daphnia* sp. immobilization and reproduction test. No. 202 OECD Guidelines for Testing Chemicals, Organization Economic Cooperation and Development, Geneva.
11. USEPA. 1985. Methods for measuring the acute toxicity of freshwater and marine organisms. Peltier W.H. and C.I. Weber, Eds., EPA-600/4–85–013. U.S. Environmental Protection Agency. 216 p.
12. Espiritu, E.Q., personal communication.

CHAPTER 6

Cyst-Based Toxicity Tests. V. Development and Critical Evaluation of Standardized Toxicity Tests with the Brine Shrimp *Artemia* (Anostraca, Crustacea)

Marleen Van Steertegem and Guido Persoone

ABSTRACT

The use of cryptobiotic stages (called cysts or dormant eggs), from which live test organisms can be hatched at will, eliminates the need for continuous stock culturing and opens a promising perspective for better standardization of test methodologies in aquatic toxicology. Although there are several groups of aquatic invertebrates which produce resting stages, the brine shrimp *Artemia* (Crustacea, Anostraca) was the very first organism used in toxicity tests starting from cysts.

0-87371-845-3/93/$0.00 + $0.50

Considering the need for simple but at the same time reliable test methods for toxicity screening in estuarine and coastal environments, a number of standard methods have been developed, based on both lethal and sublethal endpoints, and with different life stages of the brine shrimp. Since there are no specific guidelines for the development of a standard toxicity test, a general framework was worked out for this specific goal. Next to tests with the conventional criteria involving mortality, growth, and reproduction, standard protocols were worked out with less conventional test criteria, e.g., hatchability, phototactic behaviour, food uptake, and recovery.

Taking into account the importance of costs-effectiveness for routine testing, the standard acute mortality test with *Artemia* larvae has recently been modulated in a "kit", containing all (disposable) material to perform a number of toxicity tests. Based on the results of a large intercalibration exercise in Europe, the U.S., and Canada, the test methodology has been further improved and the *Artemia* cyst-based toxicity test is now available under the name of Artoxkit for cost-effective routine screening of xenobiotics in estuarine and marine environments.

KEY WORDS

Artemia, toxicity tests, standardization, cysts.

INTRODUCTION: CYST-BASED TOXICITY TESTS

Like all analytical methods prescribed for routine use, standardization of ecotoxicological tests is highly desirable for reasons of uniformity and reproducibility; furthermore, it is mostly a prerequisite for the use of such tests in a regulatory framework.[1,2] Scientific literature on ecological testing contains a great number of methods, ranging from (rather) simple short-term tests on single species, to (very) complex long-term multispecies tests. Whereas in human toxicology, all bioassays are standardized to a very high degree and make use of genetically well-defined "lines" of a few test species (such as, e.g., mice and rats), this is much less the case in terrestrial as well as in aquatic ecotoxicology. The present lack of standardization in ecotoxicological testing is such that different laboratories can produce widely varying test results on the same test material while still in technical compliance with the method.[3]

In 1983,[4] Maki emphasized that most intercalibration exercises in ecotoxicology failed because the experimental protocols were inadequate or because the basic ecological requirements of the test species were not sufficiently taken into account.

In 1981, the U.S. National Academy of Sciences established a well-defined set of criteria for evaluating test procedures. One of these stipulates that it should be possible to standardize the test procedure to minimize the variability and to ensure data compatibility among laboratories.[2] Nevertheless, to date the need for standardized test methods remains high, not least for toxicity screening in the marine environment.

One of the major reasons for the lack of standardization in ecotoxicology is due to the fact that the bioassays are dependent to a large extent on the recruitment from the wild of (genetically uncharacterized) populations of test organisms, and/or to the different maintenance and culturing conditions of laboratory reared test species. Cowgill[5,6] stated that differences in the strain of the test species used in round robin testing are often likely to be responsible for the substantial interlaboratory variabilities in test results between the different participating laboratories.

The use of resting stages (cysts, dormant eggs) of particular aquatic species as "starting biological material" has been shown to be an attractive and practical way to bypass the problems of recruitment and/or maintenance of live stocks of test organisms which are in good health and available in sufficient numbers. "Cyst-based" toxicity tests also contribute to a higher degree of standardization since they eliminate differences in sensitivity due to variation in culturing conditions of live stocks.[7–10] Furthermore, reference strains or batches of cysts of specific test species with known characteristics can be produced and/or stored in selected laboratories from which they can be dispatched to the users without any of the precautions needed for shipment of live organisms. Last but not least, the use of resting stages also eliminates the uncertainty (if not the nightmare) about test animal availability.

The use of resting stages also has a substantial positive repercussion on the cost of the tests. Persoone and Van de Vel calculated that the recruitment and maintenance of live stocks, alternatively the continuous culturing of test organisms, can make up more than the half of the total costs for both acute and chronic toxicity tests.[11] In analogy, Kszos and Stewart concluded that cultures of *Ceriodaphnia dubia* and *Pimephales promelas* used in short-chronic tests can be cost-effective only if a laboratory conducts several tests per month.[12]

Despite the inherent advantages of the use of cryptobiotic stages, few attempts have been made so far to develop standardized toxicity tests departing from "dormant eggs" of a particular species.

One of the major research objectives of the Laboratory for Biological Research in Aquatic Pollution at the University of Ghent in Belgium over the last 15 years has been to develop "cyst-based" toxicity tests with different aquatic invertebrates species, for use as screening methods in routine aquatic toxicology. The present paper reviews the state of the art of the tests that have been worked out for the brine shrimp *Artemia* (Anostraca, Crustacea), based on various endpoints and life-stages.

ARTEMIA AS TEST ORGANISM IN TOXICITY STUDIES

Although there are several groups of aquatic invertebrates which produce resting stages, the brine shrimp *Artemia* was historically the very first to be used in toxicity tests starting from cysts.

The brine shrimp *Artemia* is a small anostracan crustacean which is commonly found in highly saline biotopes (salt lakes and salinas). Under particular (stressing) environmental circumstances, this euryhaline species shifts its reproduction from ovoviviparity (life offspring) to oviparity (cyst production), by producing embryos that are set free in a "dormant stage" and protected by a tough, resistant shell. Because of low competitive and predation pressure, *Artemia* populations can develop to considerable extent in their specific saline biotopes and produce substantial amounts of cysts.

These "dormant" embryos can be stored dry for long periods of time (up to several years) without losing their viability. When hydrated and incubated under appropriate conditions in seawater, *Artemia* cysts hatch within 24 h, giving birth to free-swimming larvae (nauplii) (Figure 1). Contrary to other organisms that also produce resting stages, the cysts of *Artemia* can be collected (in huge amounts) from natural biotopes, in a monospecific way. However, the brine shrimp can also easily be cultured in the laboratory, even on non-living (inert) food.

The advantages of dry storage and easy hatching of the cysts have led over the last 40 years to the worldwide use of this biological material as a most interesting source of "live food" for crustacean and fish aquaculture. For the same reasons, brine shrimp, have been and are still

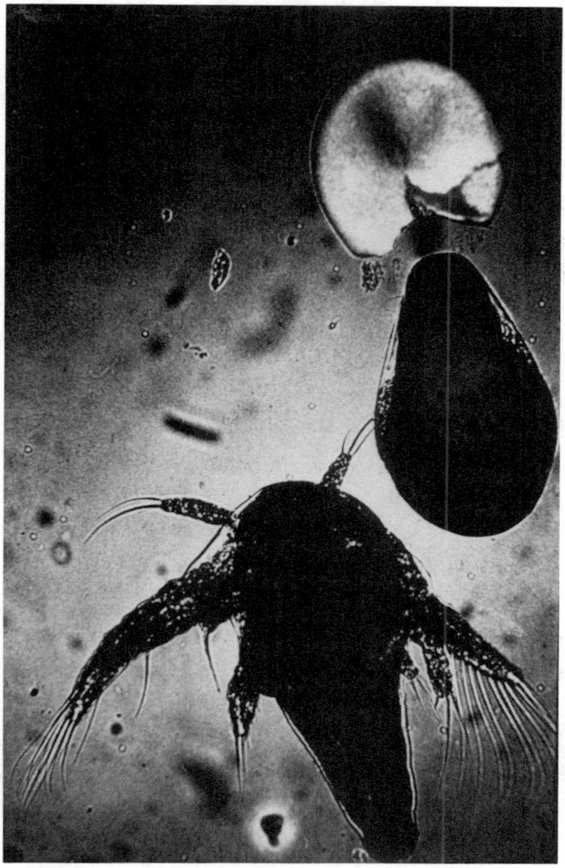

FIGURE 1. Pre-nauplius in E-2 stage and freshly hatched instar I nauplius of the brine shrimp *Artemia*.

extensively used as a most convenient material for fundamental research, encompassing virtually all biological disciplines. The bibliography on *Artemia* compiled by McCourt and Lavens includes over 3000 titles of publications devoted exclusively to research and use of brine shrimp in science and industry.[13]

The permanent availability of cysts has also triggered the use of *Artemia* as a very convenient test organism for aquatic toxicological investigations, and in the last two decades the brine shrimp has been used more and more frequently in toxicity studies. Besides laboratories investigating environmental problems, the medical, drug, and food sector are using *Artemia* quite frequently for specific assays, and over 150 papers have already been published worldwide on concentration-effect relationships of various chemicals to *Artemia*. A detailed review by Persoone and Wells describes the state of the art of *Artemia* in aquatic toxicology and discusses its future potential.[14]

The advantages of using the brine shrimp *Artemia* in toxicity testing were summarized by the latter two authors as follows:

- universal, year-round availability
- minimum maintenance (no need for stock culturing)
- low cost of the biological material
- intermediate sensitivity
- alternative species for other crustacean test species
- valuable reference organism for other test species

An additional advantage of *Artemia* is that a wealth of fundamental data on all life stages is available, which is very helpful in the research on the responses of the brine shrimp to toxicants.

An analysis of the effect data contained in the many (eco)toxicological papers dealing with *Artemia* reveals, however, great differences in sensitivity for the chemicals tested out as well as a rather poor precision (intra- as well as inter-laboratory variability) of the assays. Vanhaecke et al. concluded that none of the operational procedures described until then were detailed or stringent enough for use as a standard test.[15] This triggered the first attempts of the same authors to develop a "short-term" standard test (24-h LC_{50}) with brine shrimp larvae; this acute test was named the ARC-test, since it had been worked out in parallel with the aquacultural research on *Artemia* in the Artemia Reference Centre (ARC) at the University of Ghent in Belgium.[15, 16] In 1981, the standard ARC-test was submitted to a major intercalibration exercise involving approximately 60 laboratories; the outcome of this round-robin testing indicated that the precision of this simple and cost-effective bioassay was satisfactory both at the intra- and inter-laboratory level, since it was at least as good as or even better than that of the standard acute *Daphnia* and fish tests.[17–19]

Since the time of development of the ARC-test, extensive research has been performed in the same laboratory for several years on the development of standard *Artemia* toxicity tests with various life-stages and endpoints. During the course of these investigations, the influence of both biotic and abiotic factors was determined on the concentration-effect relationship of xenobiotics to *Artemia*, in order to adapt the test procedures to specific purposes. At the same time, the sensitivity of the brine shrimp was determined for different chemical products and compared with that of other test organisms currently used in ecotoxicology. These investigations presently allow us to make a critical evaluation of the usefulness of *Artemia* in aquatic toxicology, with identification of the possibilities as well as of the limitations of *Artemia* tests, as detailed hereafter.

DEVELOPMENT AND EVALUATION OF STANDARDIZED TEST METHODS WITH THE BRINE SHRIMP *ARTEMIA*

Since no specific guidelines are described in scientific literature for the development of a standard toxicity test, a general framework for this specific goal was worked out. Developing a methodology for a test starting from cysts does not differ in principle from the approach of a bioassay method with test organisms taken from live stock cultures. The part of the protocol dealing with the culturing of the animals (with all inherent biological and technical problems) is simply replaced by a specific description of the hatching procedure of the cysts; the toxicity test proper, in turn, is analogous in both cases.

Table 1 lists the different steps that have been followed in a tiered way during the development and the subsequent evaluation of standardized test methods for routine ecotoxicological testing. It is clear that this scheme which has been worked out for *Artemia* can also be applied to other test organisms and not only for cyst-based tests but also for other toxicity tests. If followed, the outcome of the tiered process should be a detailed and waterproof test with a precise description of the experimental conditions in which sufficient attention is paid to all the parameters that may influence the results and the precision (both repeatability and reproducibility) of the test.

The first step, viz. the "development and standardization of the test methodology", implies a detailed study of biological, chemical, physical, and technological factors. The following parameters which influence the variability of the test results and the sensitivity of the test organisms, were taken into consideration and extensive experimentation was carried out to determine "optimal" test conditions:

Table 1
Stepwise Procedure in the Development and Evaluation of a Standardized Toxicity Test

1. Development and standardization of the test methodology
   - Selection of test species
   - Selection of test criterion and test duration
   - Analysis of biotic parameters and selection of standard values
   - Analysis of abiotic parameters and selection of standard values
   - Determination of the practicability of the experimental procedure
   - Determination of the precision (repeatability) of the test
   - Selection of reference strain of the test species
   - Selection of reference chemical for quality control tests
2. Evaluation of the usefulness of the developed test method
   - Determination of the sensitivity of the test for various types of toxicants
   - Comparison of the test with other toxicity tests in terms of sensitivity and practicability
3. Interlaboratory evaluation (round robin testing)
   - Evaluation of the precision (reproducibility) and acceptability of the test
   - Adjustment of the test procedure (back to Step 1)
4. Submission of the standard test for adaption for use at the national/international level in a regulatory context

- test criterion and test duration
- biotic parameters — geographical origin of the cysts, life stage and age of the test organisms, diet (before and possibly during the test)
- abiotic parameters — hatching and dilution medium, temperature, salinity, light, etc.
- practibility — equipment, personnel
- repeatability
- reference strain and reference chemical for quality control tests

As emphasized by Persoone,[20] the optimum conditions were frequently a compromise between the factors "sensitivity" and "ecological representativeness" on one hand, and "practicability" and "cost" on the other hand. An increase in sensitivity and ecological representativeness almost automatically will lead to increased costs and complexity of the test method.

In a first phase of the research, acute and (sub)chronic test methods were worked out based on mortality, growth, and reproduction. Besides this, the usefulness of less conventional sublethal endpoints, viz. hatchability of the cysts, phototactic behavior, food uptake, and recovery were examined. For each of the endpoints considered, a simple technique for quantitative evaluation of the test criterion was worked out and a suitable test period was determined.

These studies revealed that the following *biological parameters* needed to be carefully selected and standardized:

- the geographical origin and storage conditions of the cysts
- the age of the test organisms
- the culturing conditions, especially the diet

With regard to the first variable, it appears that *Artemia* cysts are collected worldwide and that there are a large number of geographical strains each of which may have a different sensitivity to xenobiotics. Consequently, it is imperative to have information on the origin of the cysts and their genetic characteristics. In order to avoid variability in the test results, cysts should preferably originate from one single, well-defined site. In our laboratory, certified batches of cysts are stored which can be used as reference material. The quality of the cysts is controlled on a regular basis and on request they can be sent to users all over the world. The sensitivity of the organisms can also vary with the storage conditions, so the reference cysts are stored under optimal, well-defined environmental conditions. The use of reference cysts and of a standardized methodology allows full comparison of results of toxicity tests performed at different places and at different times.

Concerning the second factor, it is well-documented that the age of the test organisms is an important variable in toxicity testing with invertebrates.[21–23] During our own research with *Artemia*, it was found that the sensitivity to lethal and sublethal concentrations of toxicants varies during the life cycle of the brine shrimp and consequently must be considered in planning toxicity tests. To avoid differences in sensitivity, the operational procedure prescribes that the test organisms must be exactly of the same age at the start of every toxicity test; therefore, the cysts have to be hatched out under strictly controlled conditions of temperature, aeration, light, and pH. Since *Artemia* larvae are more sensitive to toxicants than adults, the II-III larval stage (48–h old larvae) was selected as the most convenient for the ARC-test. Besides the higher sensitivity of this compared to instar I-larvae, more reproducible results were also obtained with it.[15] An additional but very important advantage of using very young larval stages is the fact that it is not necessary to feed the test animals prior and during the (acute) toxicity test.

The use of organisms hatched from cysts is, however, no obstacle for chronic testing with the brine shrimp; one can culture the nauplii hatched from cysts to a specific life stage (even up to the adults). In contrast to toxicity studies with fish, the effect of diet on the sensitivity of invertebrates to xenobiotic compounds has received little attention. Some authors have, however, shown that feeding before and during the test influences the sensitivity of the test organisms and even the precision of the test results.[5,24–26] A serious (and inherent) burden in all chronic tests is that feeding during the test not only complicates the test procedure to a significant extent, but can also bias the toxicity results by adsorption of part of the toxicant on the food and its subsequent ingestion by the test animal.

When studying the effects of the diet on the sensitivity of the brine shrimp, different parameters were taken in account; e.g., quality and composition of the diet, quantity of the food, feeding schedule, live food vs. "inert" diets. The use of live microalgae has not only important financial consequences for the cost of a test, but the quality of the algae can vary substantially in time, resulting in variability in the test results. The use of a standard, commercially available inert food in turn leads to a better standardization of the test methodology while the production costs of live food (which are not negligible) are cut out completely. Several types of commercially available inert feeds have been tried for rearing *Artemia* prior or during chronic tests, provided singly or in combination and in test systems providing for continuous suspension of the inert food. Suitable feeding schedules were determined ensuring good survival and growth of the brine shrimp.

The influence of *abiotic parameters* on the toxic effects of chemicals has been studied extensively and many authors have stressed that physicochemical parameters have to be well-defined and controlled to obtain reproducible results.[5,6,27] In a recent study on the toxicity of liquid waste from the titanium dioxide industry to marine biota, it was found that the origin and hence the composition of the seawater used as dilution medium was a major determinant in the variability of toxicity results with both larvae and adults of the brine shrimp.[28] For reasons of both standardization and provisioning, the use of a synthetic seawater with a well-known and constant composition seems to be the best choice as dilution medium for *Artemia* tests.

Most laboratory toxicity tests are short-term assays measuring the acute lethal effects of the toxicants in controlled and constant conditions. Since such data do not reflect the real conditions in the environment, factorial experiments have been carried out to determine the combined effect of various temperature and salinity conditions on the sensitivity of brine shrimp larvae for two reference chemicals — sodium laurysulphate and potassium dichromate.[29] The results revealed that the variation in toxicity resulting from changing abiotic conditions is chemically dependent and can exceed one order of magnitude.

With regard to *"practicability"*, a standard test has to be carried out in routine, has to be easy to perform, and should not require personnel with a very high level of expertise nor sophisticated equipment. Furthermore, the time schedule for the tests should fit a normal work schedule. The practicability of a test should, however, not be detrimental to its precision; a standard toxicity test automatically implies an acceptable degree of precision at both the intra- and inter-laboratory level.

Regular testing with a *reference chemical* should be a mandatory exercise in routine investigations for quality control of the test procedure with regard to both the abiotic and the biotic variability.

With the aid of the tiered process mentioned above, a series of standardized test methods was developed with *Artemia*, using different life stages and endpoints. A detailed description of the standard operational procedures of these is reported elsewhere. Each method is described in a simple, stepwise, and easy-to-follow protocol, including detailed procedures for quality control, data analysis, and criteria for valid testing. Table 2 gives an overview of the main characteristics of the various methods developed, with an indication of the age of the test organisms, the test critera, and the test period. The sensitivity of the different endpoints for two reference chemicals, viz. sodium laurylsulphate (SLS) and potassium dichromate ($K_2Cr_2O_7$), is summarized in Table 3. It can be noted that the variation in sensitivity of the different test criteria is almost a factor 100 for SLS and more than a factor of 400 for $K_2Cr_2O_7$.

A very important step in the preparation of a standard test is the interlaboratory evaluation of the operational procedure to determine the degree of reproducibility. On the basis of the data and the comments gathered, the test protocol can, if necessary, be adjusted and improved by reconsidering some of the factors outlined in Step 1 of Table 1.

The final step for a good standard toxicity test is its "national and international acceptance" for legal purposes. However, this is no longer in the hands of the scientists, but in those of the decision makers and the

Table 2
Synthesis of Test Conditions for the Different Standardized *Artemia* Toxicity Tests Developed by the Authors

Type of test	Age of test organisms at start of the test	Effect measured	Test duration
Acute mortality test with larvae (ARC-test)	48 h old (instar II-III)	Mortality	24 h
Acute mortality test with adults	2–3 weeks	Mortality	24–96 h
Growth test	24 h old (instar I)	Length	7 d
Reproduction test	Preadults (± 10 d old)	Number of offspring/female/day	2–3 weeks
Hatching test	Cysts	Hatching success (free-swimming larvae)	48 h
Phototactic behavior test	48 h old (instar II-III)	Migration towards light	2.30 h
Food uptake test	48 h old (instar II-III)	Feeding rate	4 h
Recovery test	Larvae (after ARC test)	Mortality, after transfer into fresh medium	24 h

Table 3
Sensitivity of the Different Endpoints for the Reference Chemicals Sodium Laurylsulphate (SLS) and Potassium Dichromate ($K_2Cr_2O_7$)

Type of test	Test criterium	SLS ($mg.l^{-1}$)	$K_2Cr_2O_7$($mg.l^{-1}$)
Acute mortality test with larvae	LC50-24 h	19	35
Acute mortality test with adults	LC50-24 h	76	100
Growth test	LOEC-7 d	20	30
Reproduction test	LOEC-3 weeks	10	1
Hatching test	EC50-48 h	41	420
Phototactic behavior test	LOEC-4 h	1	50
Food uptake test	LOEC-2.5 h	15	30
Recovery test	Mortality at LC10 of acute test with larvae	50% mortality	100% mortality

authorities. Over the last decades, history has repeatedly shown that a standardized test is not easily accepted by the regulators. Quite often, the only reason for this is that the test is not very well known, nor frequently used. This creates a vicious cycle since the routine use of a toxicity test is usually coupled with its acceptance and imposition by the legislators.

THE ARTOXKIT, A COST-EFFECTIVE SCREENING TEST FOR ESTUARINE AND MARINE ENVIRONMENTS

Taking into account cost-effectiveness as one of the most important criteria for routine screening tests, the acute toxicity test with brine shrimp larvae (ARC-test) has recently been developed into a "kit", in analogy to the present trends in chemistry and medicine. The so-called "Artoxkit" contains (besides the cysts) all (disposable) materials necessary to perform 6 complete acute toxicity tests with a minimum of laboratory equipment. It is evident that besides the advantages of simplicity and cost-effectiveness of these cyst-based tests, a second major asset in the Toxkit concept is the high degree of standardization resulting from the use of identical biological material (standard cysts) and identical test materials (test container and medium), by all the performers of the test.

After extensive "in-house" testing for repeatability, a large Toxkit international intercalibration exercise was carried out in 1989 in Europe, the U.S., and Canada including the *Artemia* larvae test with the prototype of the Artoxkit. Copper sulphate was used as the reference chemical. A detailed report of the organization and the outcome of this round-robin testing, involving 129 laboratories for the *Artemia* test, is described in Reference 31. A summary of the results of the Artoxkit intercalibration is given in Table 4 with an indication of the number of successful/unsuccessful tests and the mean LC_{50}s (mg/l Cu^{2+}) with the coefficient of variation. Although the outcome of the intercalibration exercises was satisfactory, the overall precision was not impressive (C.V.: 50% in Europe and the U.S., 33% in Canada). Part of this variability was undoubtedly due to the fact that some participants were not at all familiar with ecotoxicological tests and that even the others were confronted with an unknown and new methodology, which requires some micropipeting skills. Another source of variation was probably due to the "crudeness" of the preparation of the toxicant dilutions in plastic test tubes with the aid of plastic pipets. The intra-laboratory variability of Artoxkit data in the laboratory of the organizers is indeed 10 to 15% when the toxicant dilutions are prepared with calibrated glass pipets in conventional analytical glassware. Nevertheless, it was undoubtedly the selection of copper sulphate as a reference toxicant which was the source of the largest part of the variability because of problems with instability of this chemical in seawater and difficulties in defining

Table 4
Results of the Artoxkit Intercalibration Exercise

	Europe	U.S.	Canada	Overall
Results received	81	25	23	129
Successful tests	69	24	19	112
Control mortality	0	0	0	0
Hatching problems	0	0	0	0
Other problems	12	1	4	17
Mean LC_{50} (mg/l^{-1} C^{2+})	3.5	3.2	3.8	3.2
C.V. (%)	49.8	50.7	32.8	47.1

exactly the mortality of the test organisms (dead vs. moribund). Based on the results of this intercalibration exercise and suggestions of the participants, the methodology has been substantially improved. The Artoxkit M is now commercially available, together with three other Toxkits with other test species (Rotoxkit F and M, and Streptoxkit F) for routine screening of xenobiotics in freshwater and/or estuarine-marine environments, respectively.

CONCLUSIONS

As a result of intensive research during the last decade in the laboratory of the authors, a series of lethal and sublethal standardized toxicity methods have been developed with the brine shrimp *Artemia*, based on different endpoints. Development of those has involved careful screening and selection of all abiotic and biotic test conditions. Each method is described in a detailed protocol, for routine toxicity testing of xenobiotics in estuarine and marine waters.

Considering the advantages in terms of practicability and price of the acute mortality test with *Artemia* larvae, a so-called Artoxkit was developed as a very convenient and cost-effective tool for rapid toxicity screening of chemicals and/or wastes.

All data and evidence available indicate that brine shrimp deserves a serious consideration as a valuable and very practical test organism in a battery of well-chosen tests for toxicity screening of xenobiotics in the estuarine and/or marine environment.

REFERENCES

1. Hammons, A.S. 1981. *Methods for Ecological Toxicology. A Critical Review of Laboratory Multispecies Tests.* Ann Arbor Science. Michigan.
2. National Research Council. 1981. *Testing for Effects of Chemicals on Ecosystems. A report by the Committee to Review Methods for Ecotoxicology.* National Academy Press, Washington, DC.
3. Williams, L.R. 1985. Harmonization of biological testing methodology: a performance-based approach. In R.C. Bahner and D.J. Hansen, Eds., *Aquatic Toxicology and Hazard Assessment: Eight Symposium.* ASTM STP 891, American Society for Testing and Materials, Philadelphia, pp. 288–301.
4. Maki, A. 1983. Ecotoxicology — critical needs and credibility. *Environ. Toxicol. Chem.* 2:259–260.
5. Cowgill, U.M. 1987. Critical analysis of factors affecting the sensitivity of zooplankton and the reproducibility of toxicity test results. *Water Res.* 21:1453–1462.
6. Cowgill, U.M. 1986. Why round-robin testing with zooplankton often fails to provide acceptable results. In T.M. Poston and R. Purdy, Eds., *Aquatic Toxicology and Environmental Fate: Ninth Volume.* ASTM STP 921, American Society for Testing and Materials, Philadelphia, pp. 349–356.
7. Dive, D. and G. Persoone. 1984. Protozoa as test organisms in marine ecotoxicology: luxury or necessity. In G. Persoone, E. Jaspers and C. Claus, Eds., *Ecotoxicological Testing for the Marine Environment*, Vol. 2. State University Ghent, Belgium, Inst. Mar. Sci. Res., Bredene, Belgium, pp. 281–305.
8. Persoone, G., L. Brendonck, E. Cotou, C. Janssen, A. Van de Vel and M. Van Steertegem. 1990. The use of dormant eggs (ephippia) of *Daphnia* for toxicity tests as a cost-effective alternative to continuous culturing of live stock. Commission of the European Communities, Final Report, Contract No B6614-89-44.
9. Snell, T.W. and G. Persoone. 1989. Acute toxicity bioassays using rotifers. I. A test for brackish and marine environments with *Branchionus plicatilis. Aquat. Toxicol.* 14:751–765.
10. Snell, T.W. and G. Persoone. 1989. Acute toxicity bioassays using rotifers. II. A freshwater test with *Branchionus rubens. Aquat. Toxicol.* 14:81–92.
11. Persoone, G. and A. Van de Vel. 1987. Cost-analysis of 5 current aquatic ecotoxicological tests. Report EUR 1134EN, Environment and quality of life. Commission of the European Communities, DGXI-Environment, Consumer Protection and Nuclear Safety.

12. Kszos, L.A. and A.J. Stewart. 1991. Effort-allocation analysis of the seven-day fathead minnow (*Pimephales promelas*) and (*Ceriodaphnia dubia*) toxicity tests. *Environ. Toxicol. Chem.* 10:67–72.
13. McCourt, R.P. and P. Lavens. 1985. The brine shrimp *Artemia* - Bibliography. D. Versichele, P. Lavens and P. Sorgeloos, Eds., Artemia Reference Centre, State University of Ghent, Belgium.
14. Persoone, G. and P.G. Wells. 1987. *Artemia* in aquatic toxicology: a review. In P. Sorgeloos, D.A. Bengtson, W. Decleir and E. Jaspers, Eds., *Artemia Research and its Applications*, Vol. 1-Morphology, Genetics, Strain Characterization, Toxicology. Universa Press, Wetteren, Belgium, pp. 259–275.
15. Vanhaecke, P., G. Persoone, C. Claus and P. Sorgeloos. 1980. Research on the development of a short term standard toxicity test with *Artemia* nauplii. In G. Persoone, P. Sorgeloos, O. Roels and E. Jaspers, Eds., *The Brine Shrimp Artemia*, Vol. 1- Morphology, Genetics, Radiology, Toxicology. Universa Press, Wetteren, Belgium, pp. 263–285.
16. Vanhaecke, P., G. Persoone, C. Claus and P. Sorgeloos. 1981. Proposal for a short-term toxicity test with *Artemia* nauplii. *Ecotoxicol. Safety* 5:382–387.
17. Persoone, G., P. Vanhaecke, P. Sorgeloos and C. Claus. 1981. Intercalibration exercise on a short-term standard toxicity test with *Artemia* nauplii. Commission of the European Communities. Final Report. Contract EEC ENV. 369B(N).
18. Vanhaecke, P. and G. Persoone. 1981. Report on an intercalibration exercise on a short-term standard toxicity test with *Artemia* nauplii (ARC-test). In H. Leclerc and D.Dive, Eds., *Les tests de toxicité aigue en milieu aquatique. Methodologie—Standardisation—Interprétation.* INSERM, Paris, France, pp. 359–376.
19. Vanhaecke, P. and G. Persoone. 1984. The ARC-test: a standardized short-term routine toxicity test with *Artemia* nauplii. Methodology and evaluation. In G. Persoone, E. Jaspers and C. Claus, Eds., *Ecotoxicological Testing for the Marine Environment*, Vol. 2. State University Ghent, Belgium, Inst. Mar. Sci. Res., Bredene, Belgium, pp. 143–157.
20. Persoone, G. 1980. Standardization of aquatic bioassays: compromises between biological and economical criteria. In J.F. Klaverkamp, S.L. Leonard and K.E. Marshall, Eds., *Proc. 6th Annu. Aquat. Toxicol. Workshop. Can. Tech. Rep. Fish. Aquat. Sci.* 975. pp. 111–122.
21. McCahon, C.P., A.J. Whiles and P. Pascoe. 1989. The toxicity of cadmium to different larval instars of the trichopteran larvae *Agapetus fuscipes* Curtis and the importance of life cycle information to the design of toxicity tests. *Hydrobiologia* 185:153–162.

22. Nebeker, A.V., M.A. Cains and C.M. Wise. 1984. Relative sensitivity of *Chironomus tentans* life stages to copper. *Environ. Toxicol. Chem.* 3:151–158.
23. Williams, K.A., D.W.J. Green, D. Pascoe and D.E. Gower. 1986. The acute toxicity of cadmium to different larval stages of *Chironomus riparius* (Diptera: Chironomidae) and its ecological significance for pollution regulation. *Oecologia* (Berlin) 70:362–366.
24. Pesch, C.E., P.S. Schauer and M.A. Balboni. 1986. Effect of diet on copper toxicity to *Neanthes arenaceodentata* (Annelida, Polychaeta). In T.M. Poston and R. Purdy, Eds., *Aquatic Toxicology and Environmental Fate: Ninth Volume.* ASTM STP 921, American Society for Testing and Materials, Philadelphia, pp. 369–383.
25. Taylor, M.J. 1985. Effect of diet on the sensitivity of *Daphnia magna* to linear alkylbenzene sulfonate. In R.D. Carwell, R. Purdy, and R.C. Bahner, Eds., *Aquatic Toxicology and Hazard Assessment: Seventh Symposium.* ASTM STP 854, American Society for Testing and Materials, Philadelphia, pp. 53–72.
26. Verriopoulos, G. and M. Moraitou-Apostolopoulo. 1989. Toxicity of zinc to the marine copepod *Tisbe holothuriae*; the importance of the food factor. *Arch. Hydrobiol.* 114:457–463.
27. McLusky D.S., V. Bryant and R. Campbell. 1986. The effects of temperature and salinity on the toxicity of heavy metals to marine and estuarine invertebrates. *Oceanogr. Mar. Biol. Ann. Rev.* 24:481–520.
28. Persoone, G., M. Van Steertegem and L. Fernandez Fuentes. 1990. Toxicity tests with the brine shrimp *Artemia* for waste from the titanium dioxide industry. Commission of the European Communities. Final Report. Contract No B6612-420-89.
29. Persoone, G., A. Van de Vel, M. Van Steertegem and B. De Nayer. 1989. Predictive value of laboratory tests with aquatic invertebrates: influence of experimental conditions. *Aquat. Toxicol.* 14:149–166.
30. Persoone, G. and J. Castritsi-Catharios. 1989. A simple bioassay with *Artemia* larvae to determine the acute toxicity of antifouling paints. *Water Res.* 23:893–897.
31. Persoone, G., T. Snell, C. Blaise, C. Janssen and M. Van Steertegem. 1990b. TOXKITS. Intercalibration exercise in Europe, the U.S.A. and Canada. Final Report. 27 pp.
32. Janssen, C., personal communication.

CHAPTER 7

Aquatic Toxicity Tests for Selecting Chemicals for Use Offshore

Graham F. Whale, Fiona M. Fairhurst, and Andrew E. Girling

ABSTRACT

Use and discharge of chemicals offshore in oil and gas exploration and production operations in the U.K. sector of the North Sea is currently regulated by the Department of Energy's Chemical Notification Scheme. Under this scheme, notification and consent for approval of chemicals is initially based on data for their acute toxicity to the brown shrimp *Crangon crangon*. However, in response to declarations made during the Third International Conference on the Protection of the North Sea, national regulatory schemes will be replaced by an international harmonized scheme developed by the Paris Commission to approve of use and discharge of chemicals offshore.

0-87371-845-3/93/$0.00 + $0.50

To provide relevant ecotoxicological data for a harmonized procedure for approval, the Paris Commission is seeking to have available tests for a limited number of marine species that are:

1. representative of important trophic levels
2. susceptible to relevant pollutants
3. commonly available with a wide distribution
4. amenable to laboratory testing

Currently, test methods with representative algal, herbivore, and sediment reworker species are under consideration.

In this chapter, chemicals designed for use offshore (candidate oil and gas pipeline protection packages) have been used in assessing test methods with four species of marine organisms and for comparing the relative sensitivity of this marine test with that of the standard freshwater toxicity test with *Daphnia magna*. The marine test species were the brown shrimp *C. crangon* (the traditional U.K. test species), the luminescent bacterium *Photobacterium phosphoreum* (Trade name Microtox) and two marine herbivores which are currently being considered by the Paris Commission; the calanoid copepod *Acartia tonsa*, and the oyster *Crassostrea gigas*. These tests are discussed in relation to their suitability for providing data to assess the environmental acceptability of chemicals used offshore in light of the likely changes in the regulatory requirements.

KEY WORDS

Hazard assessment, toxicity, *Photobacterium*, *Daphnia*, *Crangon*, *Crassostrea*, *Acartia*.

INTRODUCTION

At the Third International Conference for the Protection of the North Sea in March 1990, ministers representing 9 European countries and the commission of the European Communities agreed to adopt mandatory controls for the use and discharge of chemicals by the offshore oil and gas industry. In response to this declaration, the Paris Commission (PARCOM) set up an International Working Group to evaluate the chemical hazard assessment data needed to form the basis of a harmonized mandatory control scheme. In this, it is envisaged that acute aquatic toxicity data would be used to rank chemicals in terms of their potential to cause toxic effects in the marine environment. However, it is recognized that the hazard assessment of chemicals used offshore will also be based on other factors, i.e., the biodegradability and potential for bioaccumulation of the chemicals. A series of candidate marine species toxicity tests are currently being ring tested and assessed by PARCOM with a view to harmonize procedures for evaluating and testing chemicals intended for use offshore.

In this chapter, two of the PARCOM candidate toxicity tests (an acute test with the marine calanoid copepod *Acartia tonsa* and an embryo development test with the oyster *Crassostrea gigas*), and two other well documented marine tests (an acute test with the brown shrimp, *Crangon crangon* and a luminescence inhibition test with the bacterium, *Photobacterium phosphoreum*, trade name: Microtox) have been assessed for their suitability for inclusion into approval procedures for selecting chemicals for use offshore. This assessment has been based on:

1. the extent to which the test organism is able to fulfil the criteria of an "ideal" test species
2. the relative sensitivity of each test to a range of pipeline protection chemicals
3. the relative cost of conducting each toxicity test to a standard suitable for regulatory evaluation

The acute toxicity test with the freshwater cladoceran *Daphnia magna* has also been included in this evaluation for comparative purposes. This is one of the most widely conducted tests for obtaining aquatic toxicity data for chemicals and consequently a large data base exists for a wide variety of chemicals. Data from this test are currently required for notification of chemicals under the 6th Amendment to the EEC's "Dangerous Substances" Directive.

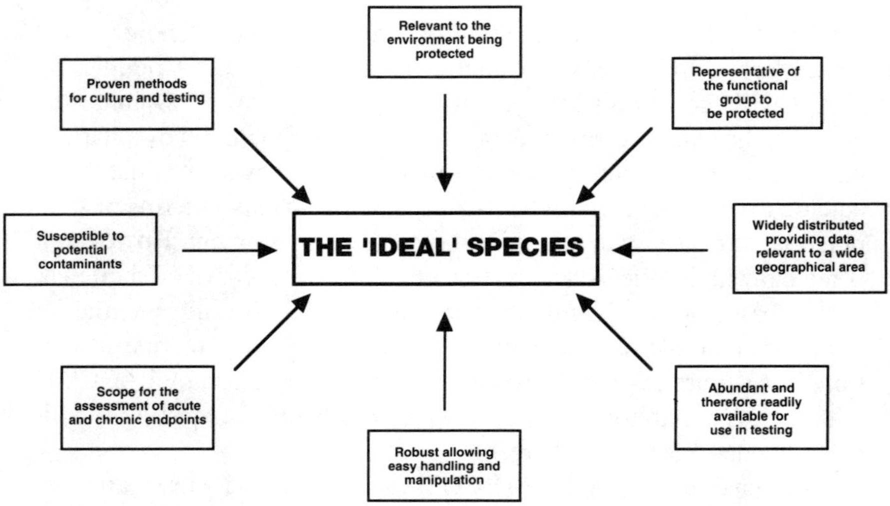

FIGURE 1. Criteria for selecting appropriate test species.

Criteria for an "Ideal" Test Species

Criteria for an ideal test species are shown in Figure 1. Proven test procedures exist for all of the species under consideration here and all are readily available for testing. A summary of how the species conform with the other selection criteria, based on our experience is represented in Table 1.

TOXICITY TEST PROCEDURES AND THEIR SUITABILITY FOR INCLUSION INTO A REGULATORY SCHEME

Daphnia magna

Daphnia magna is an important freshwater toxicity test species and internationally accepted procedures exist for both 48-h acute immobilization and chronic reproduction toxicity tests.[1–3] Under the Sixth Amendment to the EEC's "Dangerous Substances" Directive, acute and chronic *D. magna* toxicity data may be required for the hazard assessment of the chemicals at Base set, and Level 1 (1 tonne per annum and 10 tonnes per annum, respectively). As a result, a toxicity data base for a wide variety of chemicals exists for this species. The static acute

TABLE 1

Qualitative evaluation of the suitability of the five test species for inclusion in a chemical hazard assessment scheme for offshore environments.

Species	Source of test organisms	Trophic Level	Representative	Widely distributed	Scope for acute/chronic tests	Robust	Extensive data base	Relevant	Susceptible
Daphnia magna	Laboratory culture	Freshwater Primary consumer	X	X	◆	◆	◆	X	◆
Photobacterium phosphoreum (Microtox)	Commercial freeze dried reagent	Heterotrophic bacterium	●	●	X	◆	◆	◆	●
Acartia tonsa	Laboratory culture	Primary consumer	◆	◆	◆	◆	●	◆	◆
Crassostrea gigas	Commercial shell fishery	Primary consumer	◆	●	●	◆	●	●	◆
Crangon crangon	Wild caught stocks	Detritovore	◆	●	X	●	◆	◆	●

Note: 1 - This is a personal evaluation by the authors: ◆ Fulfils ● Shortfall/uncertain **X** Inadequate toxicity test with *D. magna* assessed in this paper determines the concentration of a chemical which immobilizes 50% of the young (>6 h but <24 h old at the start of the test) *D. magna* after a 48-h exposure period.

Although the *D. magna* test may not appear very appropriate for a marine hazard assessment scheme, the fact that it is well proven and that a large data base already exists suggests that it should not be automatically excluded from such a scheme. The use of toxicity test species which are not obviously appropriate to the environment under consideration is not new. For example, Microtox™ tests using the marine bacterium *Photobacterium phosphoreum* have been used for many years to monitor the toxicity of effluents being discharged into freshwater environments. Even though *D. magna* is known to be susceptible to a wide range of chemicals, one concern about including test data for this freshwater species in a marine regulatory scheme would be that the behavior and hence the toxicity of chemicals could be influenced by the differences of the water chemistry. For example, the high ionic strength and buffering capacity of seawater relative to freshwater could affect the partitioning of organic acids and the ionization of functional chemical groups. This in turn could influence the toxic effects which occur.

Photobacterium phosphoreum (Microtox™)

P. phosphoreum is a heterotrophic gram negative luminescent marine bacterium which is widely distributed in the marine environment.[4] The toxicity test with *P. phosphoreum*, marketed under the trade name of "Microtox" (Microbics, Carlsbad, California), is based upon the negative relationship between the light output of the bacteria (measured using a photometer) and the concentration of a toxic chemical. *P. phosphoreum* has advantages over other aquatic toxicity tests in that it is totally portable and the bacteria can be stored frozen.[5] The results of this test are expressed in terms of the concentration of a chemical which results in a 50% reduction in the light output of the *P. phosphoreum* after 5 or 15 min exposure.

Correlations between toxicity to *P. phosphoreum* and toxicity to fish have been demonstrated for a wide variety of chemicals.[6,7] The portability of the toxicity analyser combined with the speed with which reliable test results can be obtained has led to the increasing use of Microtox™ data as part of screening programs to evaluate the toxicity of effluent discharges. The Microtox™ test has also been used to asses the toxicity of contaminated marine sediments[8] and as a method to predict compliance of drilling muds with offshore disposal permits.[9]

Recently, methods have been developed which allow the toxicity of solids such as sediments and drilling muds to be assessed by direct contact with *P. phosphoreum*. Previously, Microtox™ toxicity assessments of solid phases have relied on testing elutriates since sediment particles would absorb the light given off by the bacteria. This problem has been overcome by exposing *P. phosphoreum* directly to sediment suspensions and then recovering the bacteria, by centrifugation, after a suitable incubation period. The activity of the bacteria in the supernatant is then measured by a standard Microtox™ analyser and used as an index of the toxicity of the solid phase test samples relative to control or untreated samples.[10]

Acartia tonsa

The calanoid copepod *Acartia tonsa*, although not originally a native species, is widely distributed and abundant in European coastal waters. *A. tonsa* is amenable to laboratory culture being readily maintained on a mixed algal diet.[11]

A. tonsa has been used as a test organism to monitor water quality by the U.S. Environmental Protection Agency (EPA) as part of their ocean disposal permit program.[12] The results of the *A. tonsa* test are expressed in terms of the concentration of a chemical which results in 50% immobilisation of *A. tonsa* (17 to 21 d old at the start of the test) during a 48-h exposure period. *A. tonsa* has been shown to be susceptible to a range of organic toxicants[13] and is suitable species for conducting chronic toxicity tests.[14–16]

Crangon crangon

The brown shrimp *C. crangon* is a native, locally abundant species in the North Sea. Although *C. crangon* can be maintained in the laboratory, it is not routinely cultured for experimental purposes because of its tendency to be cannibalistic. Supplies of adult *C. crangon* of a suitable size for toxicity tests (carapace to telson length 40 to 70 mm) are therefore usually obtained by trawling natural populations.

An area of concern which exists when using test organisms obtained from natural populations relates to the condition of the organisms when collected and their previous history of exposure to toxic substances. Significant variations in the susceptibility of *C. crangon* to reference toxicants have only been reported in tests with physically and chemically dispersed oil where agitation was used.[17] These variations were thought to be due to the condition of the shrimp, the stage of the breeding cycle or the extent and method of handling prior to testing and not related to either the season or the seawater temperature at the time the shrimp were obtained. The results discussed in this study are based on the non-agitated tests which have been used since the early 1960s by the U.K. Ministry of Agriculture Fisheries and Food (MAFF) to provide marine hazard assessment data.[18] In this test, test solutions are renewed daily and the results are expressed as the chemical concentration which results in 50% mortality of the shrimp during a 96-h exposure period.

The advantages of using *C. crangon* as a toxicity test species are first, that procedures exist for testing many types of materials and chemicals used offshore, for example drilling muds and dispersed and dissolved chemicals. Second, owing largely to past and present UK regulations, an extensive and relevant toxicity test data base already exists for this species.

Crassostrea gigas

The Pacific oyster *C. gigas* is a non-native and often non-breeding member of the fauna of European coastal waters. Its presence in the North Sea is the direct result of commercial oyster farming activities. In this test, gametes are obtained by stripping sexually mature male and female oysters. The eggs are fertilized and allowed to reach the early stages of cleavage under controlled conditions. The embryos are added to test solutions to assess any toxic effects on their subsequent development. The original test procedure assessed the development of the embryos over 48 h[19] but the incubation time has been reduced to 24 h and the test improved.[20] The results for this test are expressed as the chemical concentration which results in a 50% failure of newly fertilized oyster embryos to develop into normal "D" stage veligers over a 24-h exposure period compared to controls. The oyster embryo test differs from the other procedures described in that it is an early life stage test in which intense cellular activity and organ developments occur during the relatively short exposure period.

A significant problem which may be encountered with this test relates to the supply of oysters in spawning condition throughout the year. Although oysters can be conditioned throughout the year, this is difficult for some laboratories without natural seawater supplies and large algal culture facilities. However, because of the increasing demand for embryos for testing purposes, some commercial oyster hatcheries have begun to offer conditioned oysters at any time of the year.

COMPARATIVE SENSITIVITY OF THE FIVE TESTS

The relative sensitivity of the five tests has been assessed on the basis of results from tests with a series of pipeline protection chemicals. These chemicals are all currently used or recommended for use in the U.K. sector of the North Sea to combat corrosion of offshore oil and gas pipelines prior to commissioning. The toxicity tests were only designed to determine the EC_{50} / LC_{50} values of the chemicals to within limits separated by an order of magnitude. This was considered sufficiently precise as only large variations in the relative susceptibility of the test species would be sufficient to discriminate between species. It is also worth noting that the current U.K. chemical notification scheme only uses order of magnitude concentration steps in *C. crangon* toxicity data to categorize chemicals intended for use offshore.

The results of the toxicity assessments are summarized in Figure 2. These show that, with the exception of the Microtox™ test, there is a consistent relationship between the relative sensitivity of the tests to the pipeline protection chemicals. For example, the *C. crangon* test is always an order of magnitude less sensitive than the test with the other marine crustacean *A. tonsa*, which in turn, is only slightly less sensitive than the test with the bivalve mollusc *C. gigas*. The results from the toxicity tests with *D. magna* and *A. tonsa* indicate that these two tests have similar sensitivity to the test chemicals. The toxicity of all but two of the chemicals tested to both species were within the same order of magnitude. Similarities in the relative susceptibility of *A. tonsa* and *D. magna* have also been shown for other toxicants by Girling.[13] In these studies, one of the five organic chemicals tested was more than an order

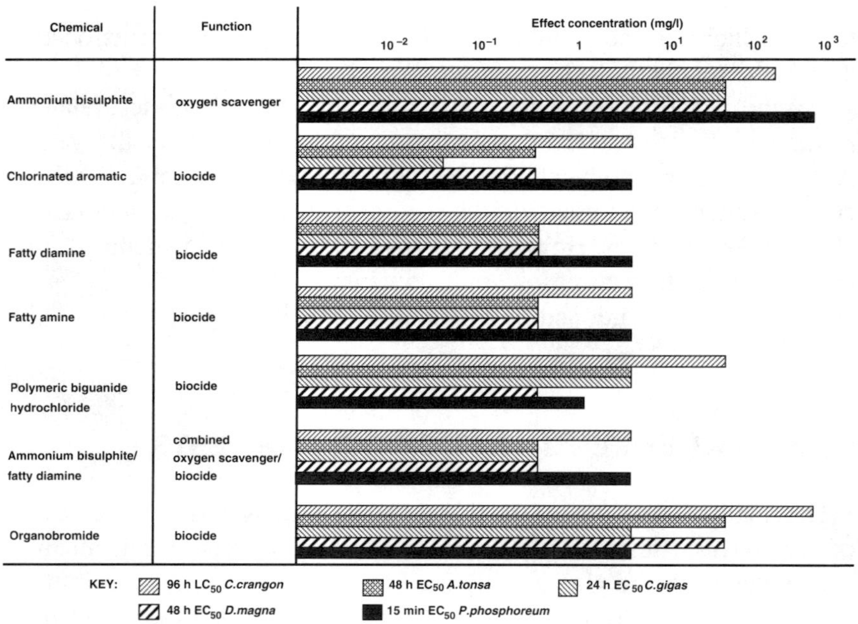

FIGURE 2. Comparative susceptibility data for the five test species to a range of pipeline protection chemicals.

of magnitude (14 times) more toxic to *D. magna* than to *A. tonsa*. However, the toxicity of the other chemicals to both species was very similar with their respective 48-h EC_{50} values differing by no more than a factor of 2.2. These initial comparisons provide some justification for including *D. magna* data, which have been previously generated in response to other regulatory requirements (e.g. EEC Directives), as part of a first tier selection procedure in a marine hazard evaluation scheme. However, it would not be sensible to rely solely on *D. magna* to categorize offshore chemicals since, under some circumstances, it may give misleading data. For example, in this study, the polymeric biguanide hydrochloride biocide was found to be more than an order of magnitude more toxic to *D. magna* than to *A. tonsa*. Use of the *D. magna* data alone would therefore result in a more harsh toxicity classification than would be justified on the basis of the *A. tonsa* data. This view is supported by data from other tests in which the polymeric biguanide hydrochloride biocide was found to be less toxic to other marine species than to *D. magna*. The converse situation applies to the chlorinated aromatic biocide since the *D. magna* toxicity data alone would result in a more favorable classification for this chemical than would be justified on the basis of the results of the *A. tonsa* toxicity test.

The results of the Microtox™ test are not consistently related to those from tests with any of the other species. This is highlighted in the more detailed comparison of the Microtox™ and *A. tonsa* results shown in Table 2.

The ratio of the Microtox™ 15 min EC_{50} values to the 48-h *A. tonsa* EC_{50} values varies between 0.08 and 23 for the range of pipeline protection chemicals tested. One problem with using Microtox™ to assess the toxicity of pipeline protection chemicals is that a number of them are recommended as biocides to prevent microbially enhanced corrosion of the inner surfaces of pipelines. Since these chemicals have been selected on the basis of their bactericidal properties, it might be anticipated that they would be more toxic to Microtox™ than the other test species. However, this does not appear to be so since most of the biocides tested in this study are more toxic to the other species than to the *P. phosphoreum* in the Microtox™ test.

Microtox™ 15 min EC_{50} values and *A. tonsa* 48-h EC_{50} values are within a factor of 23. A similar degree of variability has been found to occur between Microtox™ and another copepod species, *Nitocra spinipes*.[21]

ASSESSMENT OF EFFORT REQUIRED TO PROVIDE TOXICITY DATA FOR EACH TEST SPECIES

A factor which needs to be considered when selecting tests for incorporation into a hazard assessment scheme is their relative cost. The costs will depend upon a variety of factors including the supply of test organisms, provision of both testing environment and equipment, as well as time necessary to carry out the test.

With the exception of the Microtox™ test, the material costs for each toxicity test are difficult to calculate since these will vary between different laboratories depending upon their locations and facilities already available. For example, tests with species such as *A. tonsa* and *D. magna* require controlled light and temperature environments for culture and testing which can be achieved in many different ways.

Although cost of materials for each test should be considered, the most important cost consideration will be associated with the time required to

TABLE 2

Detailed comparison of Microtox and A. tonsa EC50 values for the pipeline production chemicals

Chemical	Function	Microtox 15 min EC_{50} (mg l^{-1})	*Acartia tonsa* 48 h EC_{50} (mg l^{-1})	Ratio of Microtox: *A. tonsa*
Ammonium bisulphite	oxygen scavenger	250 - 500	22 (14 - 36)	11 - 23
Chlorinated aromatic	biocide	0.33 (0.28 - 0.38)	0.42 (0.25 - 0.6)	0 78
Fatty diamine	biocide	3.9 (2.3 - 6.8)	0.35 (0.23 - 0.51)	1.1
Fatty amine	biocide	3.7 (2.2 - 6.4)	0.39 (0.26 - 0.54)	0.95
Polymeric biguanide hydrochloride	biocide	1.1 (0.96 - 1.2)	4.7 (2.9 - 7.1)	0.23
Ammonium bisulphite/ Fatty diamine	oxygen scavenger/ biocide mixture	2.4 (1.3 - 3.8)	0.78 (0.56 - 1.1)	3.1
Organobromide	biocide	2.5 (1.8 - 3.6)	32 (20 - 51)	0.08

95% confidence limits shown in parenthesis.

carry out each test. On the basis of experience within our laboratory, estimates of the time to carry out each of the five test procedures to regulatory standards have been calculated and these are presented in Figure 3. The estimates include components for the following:

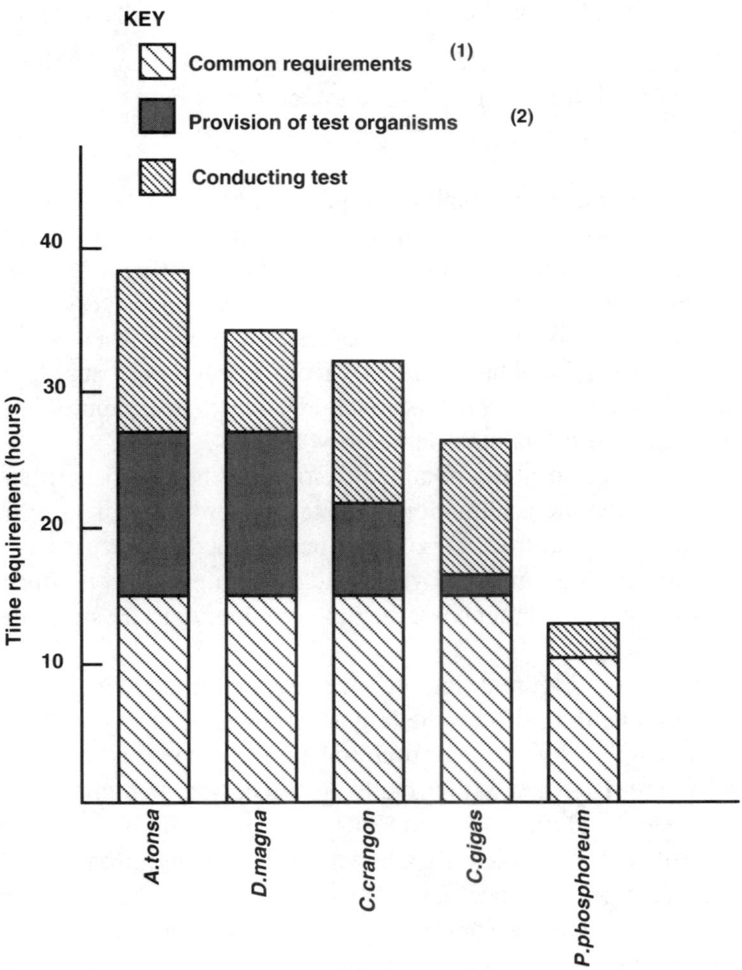

(1) Common requirements include compound control, protocol and report writing

(2) For *C.gigas* this estimate is based on oysters being obtained in spawning condition from a commercial oyster hatchery.

FIGURE 3. Comparison of time required to provide regulatory standard acute toxicity data for each of the five test species.

1. **Common requirements:** These include the time required for registering and validating the test chemical (compound control), drafting test protocols, and reporting each test.
2. **Provision of test organisms:** This includes the time required for either culturing or collecting and acclimating the test organisms.
3. **Testing:** This includes the time required for conducting both range finding and definitive toxicity tests.

For all tests there are common requirements which take up a large proportion of the total time. In all tests there are requirements for drafting of a protocol, compound control, and reporting.

The time involved in the provision of test organisms is very variable. For example, it is lowest in the case of the Microtox™ test since the *P. phosphoreum* can be obtained directly from a commercial supply when required. However, with the *A. tonsa* and *D. magna* tests, approximately 1 h/d is required to maintain laboratory cultures.

C. crangon are normally obtained from a commercial fisherman and consequently require acclimation to the testing environment prior to the start of the test. Adult *C. gigas* must be obtained either in spawning condition from a commercial oyster hatchery or must be conditioned in the laboratory. In the latter case, the time requirements are significantly greater.

It is clear from Figure 3 that the time required to conduct any of the toxicity tests is, in itself, not the largest component of any of the tests. The Microtox™ test is the least time consuming test in this respect since it is fully automated and it is normally only necessary to prepare a single chemical stock solution. Consequently, both range finding and definitive Microtox™ tests take only a few minutes. The requirements of the other tests are greater since several stock solutions of the test chemical may have to be prepared for both range finding and definitive tests and, in addition, the test concentrations in the definitive test are generally replicated and may need to be renewed. Additional time is also required for monitoring water quality parameters (i.e., temperature, pH, salinity) during the definitive test. Although both the *D. magna* and *A. tonsa* test are based on similar experimental designs, the small size and high mobility of the *A. tonsa* relative to the *D. magna* result in slightly more effort being required to set up and score the *A. tonsa* test.

In total, the Microtox™ test takes less than half the time to perform than any of the other tests. The other tests have time requirements which range from an estimated 27 h for the *C. gigas* test to 37 h for the *A. tonsa* test. It should be noted that none of these estimates includes the time required to audit the study for Good Laboratory Practice (GLP) purposes or provide analytical verification of test solution concentrations.

DISCUSSION

It is unlikely that any species will fulfill the criteria of an "ideal" species for toxicity tests in approval procedures for selection of chemicals for use offshore; individual test species and procedures will have to be judged on their relative merits.

The results presented in this chapter suggest that the two candidate PARCOM toxicity test procedures were among the most sensitive of the five examined. On this basis, either could be recommended for inclusion into a hazard assessment scheme. The experiments with the pipeline protection chemicals show that the *C. crangon* test is consistently an order of magnitude less sensitive than the *A. tonsa* test. It may therefore be possible to include the existing *C. crangon* data into a new hazard assessment scheme with an appropriate adjustment factor that takes account of this difference. The Microtox™ test requires considerably less effort than the other tests assessed but the results from this test are not consistently related to those from the other species tested. This limits the value of Microtox™ for providing toxicity data to categorize offshore chemicals. However, the Microtox™ test has advantages over some of the tests in that it can be applied to drilling muds and it could be used as a cost effective method to provide initial toxicity screening assessments. The Microtox™ test also has potential to be used as a quality control check for assessing the toxicity of approved products and as an operational means of monitoring and controlling effluent water quality.

In terms of time requirements, none of the marine tests under consideration would take significantly longer than the standard acute test with the freshwater species *D. magna*. The *C. gigas* test would take an estimated 10 h less effort than the *A. tonsa* test providing commercially conditioned oysters were used. However, *A. tonsa* results would be easier to relate to existing data from toxicity tests with other crustacean species such as *C. crangon* and *D. magna* than toxicity data for a molluscan species.

It is important that toxicity data for *D. magna* tests should not be dismissed as irrelevant in terms of their value in an approval scheme for offshore chemicals. The results from the tests with the pipeline protection chemicals presented in this chapter and those from tests with organic chemicals presented elsewhere,[13] indicate that the *D. magna* test is similar in sensitivity to the *A. tonsa* test. The similarity in sensitivity of the two tests has also been found to occur when testing a range of oil additive products.[22]

A concern which arises when extrapolating the results of toxicity tests with freshwater species to the marine environment is that, in addition to any differences in the sensitivity of the test methods which may exist, there is always the concern that differences in the chemistry of freshwater and seawater will influence the results of the tests. For instance, it is possible that the relative toxicity of the polymeric biguanide hydrochloride biocide to the freshwater and marine species was effected by the degree to which the functional groups of this biocide dissociated in the respective test media (i.e., the seawater may have inhibited the dissociation). There may be merit in PARCOM recommending national regulatory authorities to request marketing companies to also provide any data which they may have on the toxicity of their products to freshwater species when submitting their data for marine species. This would allow further comparisons to be made between the susceptibility of the freshwater and marine species and may lead to a better understanding of the chemicals which are likely to have their toxicity significantly altered due to the water chemistry of the test media. In addition to the interest in possible effects of test media on toxicity of different chemicals, requesting freshwater toxicity data for inclusion into a tiered approval scheme for offshore chemicals would be logical in view of pending EEC directives for classification and labelling of chemicals marketed in Europe.

CONCLUSION

The evaluation of the five test procedures indicates that the two candidate PARCOM toxicity tests with *A. tonsa* and *C. gigas* were the most suitable for inclusion in an approval scheme for chemicals used offshore. Both species met most of the criteria for a theoretically ideal test species, they were sensitive to the pipeline protection chemicals tested, and their test procedures were cost effective in comparison to the existing freshwater *D. magna* test.

On the other hand, the Microtox™ test did not provide data which was considered suitable for direct inclusion into a regulatory approval scheme. However, this test does have potential as a preliminary screening tool for chemicals and for monitoring changes in toxicity of, for example, batches of a chemical or effluents.

There are good reasons for not discounting toxicity data for *C. crangon* and *D. magna* from inclusion in an approval scheme. *C. crangon* and *D. magna* tests are well established as a result of U.K. and international regulations and consequently an extensive data base already exists for the toxicity of a wide range of chemicals to these species. Given the relationship (which has been demonstrated in this study) between the toxicity of pipeline production chemicals to these two species and the two PARCOM candidate species, there would appear to be some justification for accepting data from these tests as part of a first step in an approval scheme. This would avoid the need for tests to be repeated using "accepted" species and procedures.

To avoid further confusion about the acceptability of data obtained using different toxicity tests into new approval procedures, it is important that the PARCOM initiative fulfills its original objective of establishing clear guidelines for testing chemicals intended for use offshore in the North Sea. In order to achieve this, it is considered important that the PARCOM exercise does not lead to the adoption of even more toxicity test procedures. Instead, the tests should be evaluated with the aim of reducing the number of tests required to provide the most relevant toxicity data for an internationally agreed approval procedure for offshore chemicals.

REFERENCES

1. EEC. 1984. Council directive 79/831/EEC Annex 5 Part C. Methods for the determination of ecotoxicity: C.2 Acute toxicity for *Daphnia*. *Official Journal of the European Communities* No. L251/3: 155–159.
2. OECD. 1984. Guidelines for testing chemicals. Section 2: Effects on biotic systems — 202 *Daphnia* sp. Acute immobilisation test and reproduction test. ISBN-64-12221-4.

3. U.S. Environmental Protection Agency. 1975. Methods for acute toxicity testing with fish. Macroinvertebrates and amphibians. EPA/660/2-75/009.
4. Bauman, P. and L. Bauman. 1984. Photobacterium. In N.R. Krieg and J.G. Holt, Eds., *Bergey's Manual of Systematic Bacteriology*. Vol. 1. Williams and Williams, Baltimore/London, pp. 539–545.
5. Bulich, A.A.. 1979. Use of luminescent bacteria for determining toxicity in aquatic environments. In L.L. Marking and R.A. Kimerle, Eds., *Aquatic Toxicology*. ASTM STP 667. American Society for Testing and Materials, Philadelphia, pp. 98–106.
6. Bulich, A.A., M.W. Greene and D.C. Isenberg. 1981. Reliability of the bacterial luminescence assay for determination of the toxicity of pure compounds and complex effluents. In D.R. Branson and K.L. Dickson, Eds., *Aquatic Toxicology and Hazard Assessment: Fourth Conference,* ASTM STP 737. pp. 338–347.
7. Qureshi, A.A., K.W. Flood, S.R. Thompson, S.M. Janhurst, C.S. Inniss and D.A. Rokosh. 1982. Comparison of a luminescent bacterial test with other bioassays for determining toxicity of pure compounds and complex effluents. In J.G. Pearson, R.B. Foster and W.E. Bishop, Eds., *Aquatic Toxicology and Hazard Assessment: Fifth Conference,* ASTM STP 766, pp. 179–185.
8. Schiewe, M.H., E.G. Hawk, D.I. Actor and M.M. Krahn. 1985. Use of a bacterial luminescence assay to assess toxicity of contaminated marine sediments. *Can. J. Fish. Aquat. Sci.,* 42, 1244–1248.
9. Dorn, P.B., J.C. Raia, S.H. Evans, R.A. Balderas and E.Y. Chai. 1990. Rapid methods for predicting compliance to drilling mud toxicity tests with mysids for offshore NPDES general permits. *Environ. Auditor.*, 1, 221–227.
10. Brouwer, H., T. Murphy and L. McArdle. 1990. A sediment contact bioassay with *Photobacterium phosphoreum. Environ. Toxicol. Chem.* 9, 1353–1358.
11. Sosnowki, S.L. and J.H. Gentile. 1978. Toxicological comparison of natural and cultured populations of *Acartia tonsa* to Cadmium, Copper and Mercury. *J. Fish. Res. Board Can.*, 35, 1366–1369.
12. U.S. Environmental Protection Agency 1976. Bioassay procedures for the ocean disposal permit program. EPA-600/9-76-010, pp. 31–49.
13. Girling, A.E. 1988. An acute toxicity test with marine copepod — *Acartia tonsa*. Proceedings of the First European Conference on Ecotoxicology. Oct 17–19, 1988, Copenhagen, Denmark. pp.151–166.
14. Ward, T.J., E.D. Rider and D.A. Drozdowski. 1979. A chronic toxicity test with the marine copepod *Acartia tonsa*. Proceedings of the Second Annual Symposium on Aquatic Toxicology. ASTM Special Technical Publication 667. L.L. Marking and R.A. Kimerle (Eds.), pp.148–158.

15. Girling, A.E. 1989. Acute and chronic toxicity of produced water from a North Sea oil platform to the calanoid copepod *Acartia tonsa*. *Bull. Environ. Contam. Toxicol.* 43:280–286.
16. Sunda, W.G., P.A. Tester and S.A. Huntsman. 1987. Effects of cupric and zinc ion activities on the survival and reproduction of marine copepods. *Mar. Biol.* 94(2), 203–210.
17. Norton, M.G. and F.L. Franklin. 1980. Research into toxicity evaluation and control criteria of oil dispersants. *Fish. Res., Lowestoft,* (57).
18. Franklin, F.L. 1980. Assessing the toxicity of industrial wastes, with particular reference to variations in the sensitivity of test animals. *Fish. Res. Tech. Rep.,* MAFF *Direct. Fish. Res., Lowestoft*, (61).
19. Woelke, C.E. 1972. Development of a receiving water quality bioassay criterion based on the 48-hour Pacific oyster (*Crassostrea gigas*) embryo. *Tech. Rep. Dept. Fish. Wash.*, 9, 1–93.
20. Thain, J.E. and J.Watts. 1987. The use of a bioassay to measure changes in the water quality associated with a bloom of *Gyrodinium aureolum* Hulbert. *Rapp. P.-V Reun. Cons. Int. Explor. Mer* 187: 103–107.
21. Tarkpea, M., M. Hansson and B. Samuelsson. 1986. Comparison of the Microtox™ test with the 96-hr LC_{50} test for the harpacticoid *Nitocra sinipes*. *Ecotox. Env. Safety,* 11, 127–143.
22. Girling, A.E., unpublished data.

CHAPTER 8

Development of *In Vitro* Cytotoxicology Tests Using Fish Cell Lines

J.V. Tarazona, M. Cebrian, and A. Castaño

ABSTRACT

Here we explore the use of fish cells in culture for *in vitro* cytotoxicology assays of chemicals as a possible alternative to acute toxicity testing using laboratory animals. The results of a conventional short-term cytotoxicity test with RTG-2 cells for several chemicals and their reproducibility for intra- and inter-assay replicates are presented. Although the most important need for a wide-range use of *in vitro* cytotoxicity tests is the standardization of test protocols, the short-term test with RTG-2 cells exhibits good reproducibility.

0-87371-845-3/93/$0.00 + $0.50

INTRODUCTION

The assessment of the risk of chemicals to aquatic organisms and the total environment is a complex and difficult task which always begins by performing some toxicity tests that can identify hazards and quantify harmful effects of the chemical.

A large number of single-species toxicity tests have been described using many taxa of ecological interest. Some of them have been standardized by international organizations such as OCDE, ASTM, APHA, or the European Community. However none of these individual tests is able to provide enough information to be used as the single source of data for the ecotoxicological assessment of the risk associated with the use of a particular chemical or even to define the quality criteria or permissible effluent discharge rates. Therefore, batteries of several toxicity tests have to be used even for the preliminary steps of permissible and control strategies.

The use of *in vitro* systems for toxicological research has several advantages:[1]

1. control of environmental conditions
2. elimination of interactive systematic effects
3. reduced variability between experiments
4. simultaneous or repeated sampling over time
5. small quantities of test chemicals that are needed for complete dose-response studies
6. often cheaper and quicker
7. generate smaller quantities of toxic wastes

When cell lines instead of primary cultures are used, tests are performed without using a single live animal.

The use of fish cells for *in vitro* cytotoxicity testing started in the late 1960s[2,3] and was reconsidered in the late 1970s to the early 1980s.[4–7] Several results have illustrated the usefulness of this methodology in various aspects of aquatic toxicology; however, their use has not been expanded, and the different methodological approaches have not been standardized.

The most important question is the extent to which we can extrapolate from *in vitro* systems to live animals. However, even more problematic for aquatic ecotoxicologists is how to extrapolate from laboratory experiments with individuals to community effects. Because good correlations between toxicity on cell cultures and fish have been reported,[8,9] toxicity tests with fish cell lines present themselves as good potential alternatives to laboratory experiments with individuals.

This work presents results illustrating the different applications of fish cell lines in aquatic ecotoxicology under development in our Department. These applications include the optimization and standardization of a short-term toxicity test with RTG-2, their capability for screening structure-activity relationship studies, the combination of cytotoxicity test with separative analytical techniques like HPLC, the capacity of fish cell lines to be used for the study of sublethal effects (even those as specific and useful as the induction of cytochrome P-450 activities), and finally, their use in experimentally manipulated methodologies to determine toxicity mechanisms or relationships that can not be studied in *in vivo* models.

DESCRIPTION OF CELL LINES AND TECHNIQUES

Fish Cell Lines

The fish cell line RTG-2 has been used. RTG-2 is a fibroblastic cell line derived from rainbow trout gonads. It is commercially available.

Cells were grown in Eagle's minimum essential medium (Flow, UK) with Earle's salts, non-essential aminoacids, 200 IU penicillin/ml, 100 μg streptomicin/ml, and 0.85 g sodium bicarbonate/l, supplemented with 1.25 μg fungizone/ml, 2mM glutamine, and 10% foetal bovine serum (Flow, UK). Cells were grown at 20°C in a 5% CO_2 atmosphere.

Short Term Toxicity Test

The basic conditions of the short-term cytotoxicity test using RTG-2 cells and HPLC separations have been described elsewhere.[10,11] This test is the result of an optimization-standardization research project and will be described in detail elsewhere. Briefly, after 48 to 72 h of growth, cells are exposed for 48 h, in 96 well microtiter plates, to different concentrations of the toxic chemical in the same culture medium as above but with 1% instead of 10% of foetal calf serum. After exposure, three toxicity parameters are analyzed: intracellular content of ATP, neutral red uptake, and total proteins. Toxicity is assessed as percentage of reduction against toxicant-free control cells.

Cadmium chloride toxicity tests were performed on three different days, with six replicates in each test in order to assess the intra-assay and inter-assay reproducibility of the test. A minimum of six replicates were performed for the other chemicals tested.

In Vitro Relationship Between Copper Toxicity and Copper Concentration Using RTG-2 Cells

Copper toxicity on RTG-2 cells was assayed after 24 h of exposure using the culture medium described above or the same medium plus the addition of 40 g/l of bovine serum albumin fraction V (Sigma, Germany) or 2mM EDTA (Merck, Germany). Cells were rinsed and collected in PBS, sonicated and centrifugated at 100.000 x g. The copper content of the pellet organelles was analyzed by atomic absorption spectrophotometry (model 3030B Perkin Elmer, U.S.).

RESULTS

The results of a conventional short-term cytotoxicity test with RTG-2 are showed in Figure 1; three different EC_{50} values were obtained for each chemical, depending on the toxicity parameter to be considered. For lead, intracellular ATP appeared to be much more sensitive than the other parameters.

Figures 2 and 3 show the reproducibility for intra- and inter- assay replicates, respectively, obtained for cadmium: the EC_{50} for neutral red absorption was around 6 ppm and all values ranged between 5.3 and 6.5 ppm.

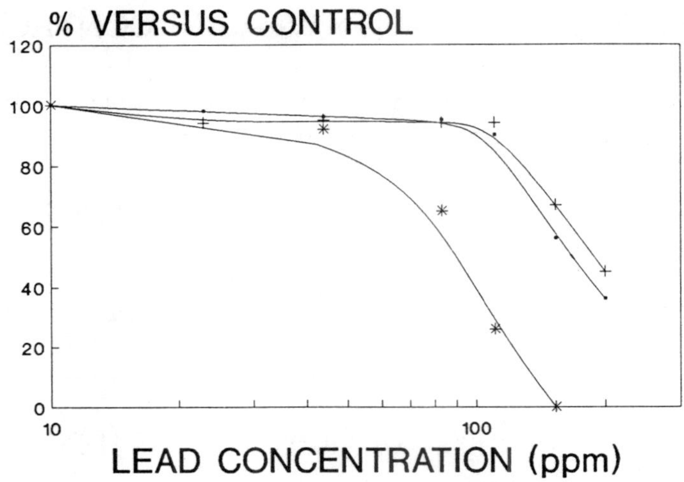

FIGURE 1. Cytotoxicity of lead on RTG-2 cells using the short-term toxicity protocol. Three different toxicity methods are used in each assay, neutral red (-.-) proteins - - (-+-) and intracellular ATP (-*-). In all cases toxicity is assessed as percentage of reduction vs. a control, toxicant-free, cells.

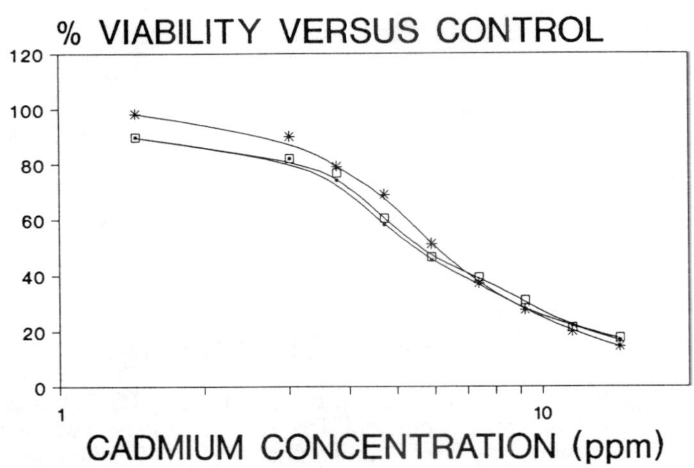

FIGURE 2. Inter-assay reproducibility of the short-term toxicity protocol with RTG-2 cells. Each curve represents the results obtained for a file of ten wells in a single 96-well-plate exposed to cadmium and total protein was used as toxicity parameter.

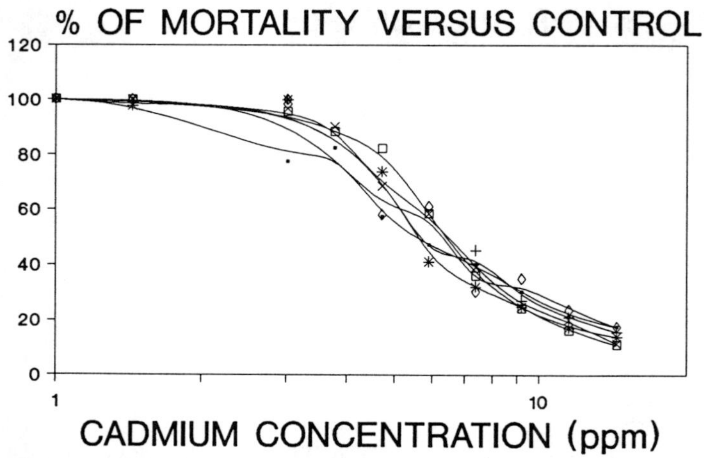

FIGURE 3. Intra-assay reproducibility of the short-term toxicity protocol with RTG-2 cells. Each curve represents the result for different plates on different days, using cadmium as chemical and total protein as toxicity parameter.

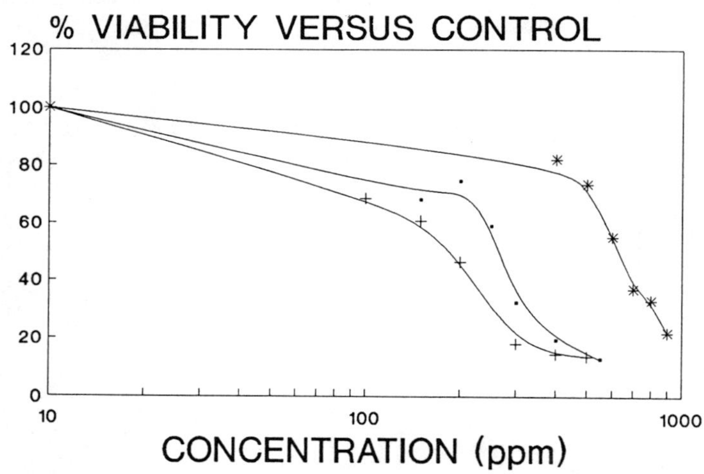

FIGURE 4. Cytotoxicity of three phenolic compounds on RTG-2 cells using the short-term toxicity protocol. Toxicity was assessed by the neutral red assay. Phenol (-*-), p-nitrophenol (-.-), 2,4 dinitrophenol (-+-).

Figure 4 shows the concentration-response curves for three related chemicals, phenol, paranitrophenol, and 2,4-dinitrophenol. An important increase in the toxicity due to the addition of nitro groups can be noted; EC_{50} values of 630, 260, and 109 ppm were recorded respectively for phenol, paranitrophenol, and 2,4-dinitrophenol.

Linear relationships were obtained for the concentration-response of copper after 24 h exposure using cell viability as the toxicity parameter. Data are described by:

$$\% \text{ cell viability} = -4.64 * [\text{copper in ppm}] + 128.38$$

with a correlation coefficient r = 0.90.

The addition of 40 g/l of bovine serum albumin to the culture medium reduced the copper toxicity by a factor slightly lower than 10, giving:

$$\% \text{ cell viability} = -0.68 * [\text{copper in ppm}] + 138.57$$
$$r = 0.99$$

While a similar relationship was obtained by the addition to the culture medium of 2mM EDTA:

$$\% \text{ cell viability} = -0.65 * [\text{copper in ppm}] + 130.23$$
$$r = 0.99$$

Initial plateaus for low copper concentrations were observed for the three test conditions. A log-linear relationship between copper concentration in cell organelles and copper toxicity was observed, independent of the addition or not of bovine serum albumin or EDTA to the culture medium (Figure 5).

DISCUSSION

Thousands of new chemicals are synthesized annually and the toxicological and ecotoxicological consequences of their use have to be predicted before their commercial use. On the other hand, millions of tons of toxic sewage are continuously dumped to aquatic environments, and its toxicity should be assessed at as short intervals as possible. In such conditions, any tool that can provide useful toxicological information should be contemplated, and the objective of this chapter is to demonstrate some aspects of the capability of fish cell lines to provide this.

The application of fish cells in culture for *in vitro* cytotoxicology assays of chemicals, as a possible alternative to acute toxicity testing using laboratory animals, has not received as much attention as that

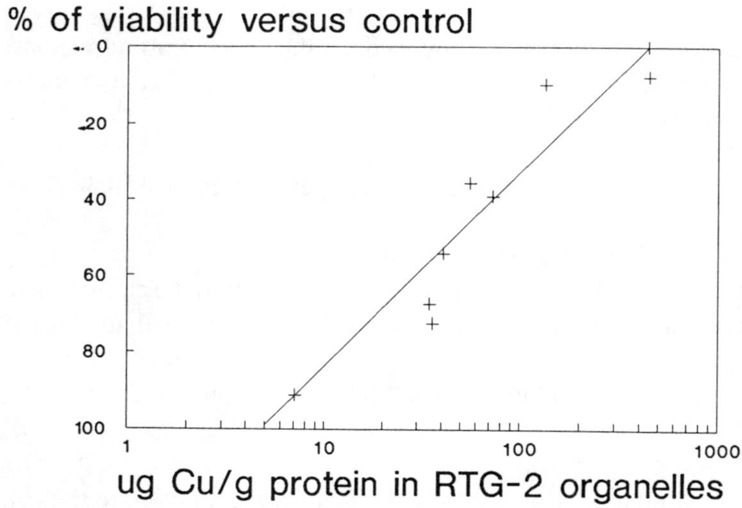

FIGURE 5. Relationship between copper content in pellet organelles and copper cytotoxicity on RTG-2 cells, assessed by the neutral red assay after 24 h of exposure. The relationship did not change by the use of culture medium rich in complexant agents.

provided to mammalian cells. However, appropriate information has been published, especially for two specific fish cell lines, RTG-2[3,5,6,8] and BF-2,[6,7,9] while other lines have also been used[9] and new fish cell lines are continuously reported.[12, 13]

Good correlation between *in vitro* EC_{50} values and *in vivo* LC_{50} values have been obtained for both RTG-2 and BF-2 cells with their related fish species,[6,8,9] and in our opinion, the most important need for the wide-range use of *in vitro* cytotoxicity test is the standardization of test protocols.

The short-term cytotoxicity test with RTG-2 cells developed in our group, exhibits the reproducibility required for a generally used toxicity test and also has other advantages. These are low cost, rapidity, versatility, and a need for only very small amounts of the toxic chemical.[1,5] In addition, our test conditions provide very good sensitivity, with EC_{50} values for cadmium, phenol, and p-nitrophenol several times lower than those previously reported for RTG-2 cells.[6,8] They are even lower for cadmium and lead than those reported for BF-2, a cell line that according to Babich et al.[7] has a greater sensitivity to pollutants than RTG-2.

Some of the potential applications of this short-term toxicity test, i.e., the screening of new chemicals or the study of structure-activity relationships, are directly derived from the low-cost high-testing-capability of the assay. These include the use of 96-well microplates and colorimetric microplate readers, used elsewhere in ecotoxicology.[14] Also, the use of this test as a biological detector of an analytical HPLC system is suitable for the very low volumes used for the assays and their compatibility with the organic solvents used in reverse-phase HPLC.[11,15]

Finally, the results in Figure 5 confirm, once again, the potential use of *in vitro* model systems that can be experimentally manipulated for use in basic research. There is a lack of relationship between copper concentration in tissues and *in vivo* copper toxicity to fish, that has been attributed to the regulation and intracellular compartmentation of this essential metal in the organisms. The use of RTG-2 cells showed very good correlations between copper concentration and toxicity; correlations that were unchanged by the addition of complexant agents to the culture medium (leading to several fold decreases of copper toxicity) and appeared at the subcellular level.

REFERENCES

1. Baksi, S.M. and J.M. Frazier. 1990. Isolated fish hepatocytes — model systems for toxicology research. *Aquatic Toxicol.* 16:229–256.
2. Rachlin, J.W. and A. Perlmutter. 1968. Fish cells in culture for study of aquatic toxicants. *Water Res.* 2:409-414.
3. Rachlin, J.W. and A. Perlmutter. 1969. Response of rainbow trout cells in culture to selected concentrations of zinc sulfate. *Prog. Fish. Cultur.* 31:94–98.
4. Kocan, R.M., M.L. Landolt and K.M. Sabo. 1979. *In vitro* toxicity of eight mutagens/carcinogens for three fish cell lines. *Bull. Environ. Contam. Toxicol.* 10:269–274.
5. Marion, M. and F. Denizeau. 1983. Rainbow trout and human cells in culture for the evaluation of the toxicity of aquatic pollutants: A study with lead. *Aquatic Toxicol.* 3:47–60.
6. Babich, H., J.A. Puerner and E. Borenfreund. 1986a. *In vitro* cytotoxicity of metals to bluegill (BF-2) cells. *Arch. Environ. Contam. Toxicol.* 15:31–37.
7. Babich, H., C. Shopsis and E. Borenfreund. 1986b. *In vitro* cytotoxicity testing of aquatic pollutants (cadmium, copper, zinc, nickel) using established fish cell lines. *Ecotoxicol. Environ. Saf.* 11:91–99.
8. Bols, N.C., S.A. Boliska, D.G. Dixon, P.V. Hodson and K.L.E. Kaiser. 1985. The use of fish cell cultures as an indication of contaminant toxicity to fish. *Aquatic Toxicol.* 6:147–155.

9. Babich, H. and E. Borenfreund. 1987. Fathead minnow FHM cells for use *in vitro* cytotoxicity assays for aquatic pollutants. *Ecotoxicol. Environ. Saf.* 14:0–10.
10. Castaño, A., B. Gallego and J.V. Tarazona. 1990. Acute toxicity of water pollutants determined by an *in vitro* cytotoxicity battery test. International Symposium on Ecotoxicology, Munich-Neuherberg, MN.
11. Tarazona, J.V., A. Castaño and B. Gallego. 1990. Detection of organic toxic pollutants in water and waste-water by liquid chromatography and *in vitro* cytotoxicity test. *Anal. Chim. Acta.* 234:193–197.
12. Moritomo, T., D.P. Anderson and W.B. Schill. 1990. Establishment of a cell line with reticulo-endothelial characteristics from a rainbow trout spleen explant. *Fish Pathol.* 25:165–170.
13. Yuanan, L., C.N. Lannan, J.S. Rohovec and J.L. Fryer. 1990. Fish cell lines: Establishment and characterization of three new cell lines from grass carp (*Ctenopharyngodon idella*). *In Vitro Cell. Dev. Biol.* 26:275–279.
14. Galgani, F. and G. Bocquene. 1991. Semi-automated colorimetric and enzymatic assays for aquatic organisms using microplate readers. *Water Res.* 25:147–150.
15. Tarazona, J.V., M.J. Muñoz, G. Carbonell, M. Carballo, J.A. Ortiz and A. Castaño. 1991. A toxicological assessment of water pollution and its relationship to aquaculture development in Algeciras Bay, Cadiz, Spain. *Arch. Environ. Contam. Toxicol.* 20:480–487.

CHAPTER 9

Replicability, Repeatability, and Reproducibility of Embryo-Larval Toxicity Tests with Fish

Göran Dave

ABSTRACT

The precision of embryo-larval (EL) fish toxicity tests has been reviewed with emphasis on their replicability (r_o), repeatability (r), and reproducibility (R). Precision, determined as CV for response variables (hatching, survival, growth) or endpoints (IC_{50}, LC_{50}, NOEC, LOEC, ZEP), decreases by approximately a factor of 2 (doubling of CV) for each step from r_o to R. The need for replication of treatments (concentrations), number of animals, and the dilution factor must be considered in relation to the desired precision, which depends on response variables and the purpose of testing. When high precision is needed (effluent

0-87371-845-3/93/$0.00 + $0.50

testing, toxicity identification evaluation, toxicity reduction evaluation) the emphasis of the experimental design should be on r_o, while in chemical screening and regulatory testing of chemicals and products the emphasis should be on R. The use of EL toxicity tests with different species of fish to estimate species-specific variability for chronic toxicity has been a neglected field of research. Furthermore, EL toxicity test methodology has a potential for assessment of effects on reproduction in wild populations of fish. When similar endpoints are used as in acute toxicity tests (EC_{50}s), the precision of EL and acute toxicity tests are equal. Therefore, species-specific differences in chronic toxicity can be determined directly without the need for application factors.

KEY WORDS

Embryo, fish, intercalibration, larvae, repeatability, reproducibility, test method, toxicity.

INTRODUCTION

Aquatic toxicology is a maturing scientific discipline. During the last decade its primary tools, the toxicity tests, have been examined for precision. Commonly used short-term tests have been intercalibrated and the use of reference toxicants has become part of the routine at laboratories using these methods. The purpose of this paper is to review the precision of toxicity tests with embryos and larvae of fish and to give some suggestions for the future productive use of these tests. To understand the present situation and to predict the future, it is essential to know about the past. Therefore, we will start with a brief historical introduction on the progress in using fish for toxicity tests with emphasis on standardization of methodology. The philosophy of using early life stages is introduced as well as the terminology used for evaluating precision.

History of Progress in Standardizing Toxicity Tests with Fish

Fish have been used in experiments for nearly 200 years, and one of the first reports of the use of fish in toxicity studies was made in 1863 by Penny and Adams.[1] In 1917, Powers[2] proposed the use of goldfish for toxicity tests and introduced the measurement of survival time vs. concentration of the toxicant. This basic concept of dose in fish, dose = f(conc., time), is still used in the application of one embryo-larval (EL) toxicity test using the median times for hatching and survival.[3,4] In 1927, Belding[5] reviewed the conditions necessary for successful testing with fish. He identified several important variables associated with the fish, such as variation in sensitivity due to individuals, species, age, size, weight and vitality (biotic factors), and environmental (abiotic) factors such as dissolved oxygen and temperature, and he stressed the importance of the biomass/volume factor in toxicity testing.

In order to define the test medium and reduce possible contamination with metals, many early investigators used distilled water as a test medium instead of natural water or tap water, but its use was terminated when the additional osmotic stress to the fish was understood.[6] Investigators using *Daphnia*, more sensitive than freshwater fish to osmotic changes,[7] introduced the first recipes for reconstituted waters prepared from distilled or deionized water and defined amounts of salts containing Na^+, K^+, Ca^{2+} and Mg^{2+} as cations and Cl^-, SO_4^{2-} and HCO_3^- as anions.[8,9]

The need for standardization of methodology increased by the number of chemicals that needed to be tested during the late 1940s when more efficient (= toxic) pesticides were introduced. One of the first variables to be standardized was the test vessel (loading factor) and the "pickle-jar era" had started, when the 5-gallon pickle jar become the standard vessel in acute toxicity tests with fish in the U.S. In 1950 and 1953, two extensive reviews on toxicity of industrial waste and its components were published by Duodoroff and Katz,[10, 11] and in the mid 1950s the first screening programs for fish toxicity were initiated at the Fish and Wildlife Service (FWS) laboratories in Leetown and Hammond Bay.[12]

In spite of the increasing number of publications on aquatic toxicology in the 1950s and 60s, standardization progressed slowly. Standardized test procedures were first published in the 11th edition of *Standard Methods for the Examination of Water and Wastewater*.[13] Then, in 1975, the Committee on the Methods for Toxicity Tests with Aquatic Organisms gave recommendations on the use of species, test temperatures, test waters, and procedures for toxicity testing.[14] Reconstituted water formulations and pH buffer schedules developed by Marking[15] were adopted by the Committee. In 1980 these procedures were adopted by ASTM (American Society for Testing and Materials) as the standard practice for conducting acute toxicity tests with fish,[16] and in the 15th (1981) edition of *Standard Methods...* many of the methods promoted by the 1975 Committee were included.[17]

In Europe, the need for regulatory testing caused a proliferation of test methods in the late 1960s and early 1970s. For determination of acute toxicity to fish, a variety of methods were used, differing in species as well as test conditions. In 1973, a "ring test" of existing methods was conducted with 11 detergents[18] and, even if almost any of the possible factors possible to standardize were released from control, the results were remarkably consistent. U.K. laboratories using 6 different methods with 4 species of fish at a water hardness from 20 to 268 mg/L expressed as $CaCO_3$ and temperatures of 15 or 20°C had an average CV for 48-h LC_{50} of 53%. Seven German laboratories testing the same 11 detergents with the German standard Golden orfe (*Idus idus melanotus* = *Leuciscus idus*) test had an average CV of 36% (calculated from Table 4[11]). Furthermore, the U.K. and the German data gave a similar ranking to the 4 least toxic and the 3 most toxic detergents, in spite of the relatively small differences in their toxicity (48-h LC_{50}s ranged from 0.73 to 7.52 mg/L). Therefore, practical concerns such as year-round and unrestricted availability of fish and political arguments related to trade regulations were also of concern in standardization of fish test methodology in Europe.

The need to standardize tests was addressed by EIFAC (European Inland Fisheries Advisory Commission) in 1975,[19] and a draft protocol for a 96-h test with zebrafish (*Brachydanio rerio*) was proposed by ISO (International Organization for Standardization) in 1976. After two intercalibrations[20,21] and a comparison with the German standard Golden orfe test[22] the proposal resulted in an ISO standard in 1984. This was divided into three parts dealing with static, semi-static and flowthrough test methods.[23] Every 5 years the ISO methods are subjected to revision, i.e., in 1991 for the fish acute test methods.

Thus, in the early 1980s several acute test methods had been subjected to intercalibrations, called ring tests in Europe and round robins in the U.S.[24] Some results of these exercises were discouraging because aquatic toxicologists realized that their standardized methods were less reproducible than they thought. There are several possible explanations for low reproducibility achieved in intercalibrations. One is that the difficulties involved in transfer of methods between laboratories is not always realized; precision is usually achieved after considerable practice. The other is too high expectations of a standard method. Today, experience has shown that standardized acute fish tests have a CV for interlaboratory precision around 40%.[21,25]

Embryo-Larval (EL) Toxicity Tests

Major reasons for using EL toxicity tests are that they are quicker and cheaper than chronic (life cycle) toxicity tests, and, more important, they can still estimate the concentration of a toxicant or mixture that is "safe" during chronic exposure. This is because early developmental stages are usually more sensitive than later stages (juveniles and adults) to stressors of various kinds (ionic composition, radiation, temperature, toxicants). A small organism (early stage) also requires less time to reach equilibrium and usually responds faster than a large organism. Therefore, EL toxicity tests can provide a reasonably precise (within a factor of 2) estimate of the chronically safe concentration.[26–28] Furthermore, for practical reasons life-cycle toxicity tests can only be performed with a limited number of fish species, while EL toxicity tests can be made with a considerable number of species and especially with native fish which require protection, on a case to case basis.

During the late 1980s, two tendencies in the development and use of EL toxicity tests emerged. The period of exposure was reduced from the originally proposed 1 or 2 months[27] to 2[3] or even 1 week(s).[29] Since eggs (embryos) are usually more resistent to chemicals than larvae, the embryo stage exposure has been omitted in some early life stage tests.[30,31] Somewhat surprisingly, the possibility of using a variety of species[32,33] has not been explored further, but instead a few cultured and/or laboratory species have been favored.

The Need for Precision and Accuracy of Test Methods

The need for precision and accuracy are dictated primarily by the intended use of the results and possibly also by their potential use in the future (databases). For screening of chemicals and products, toxicity to fish is only one of several parameters used by regulatory authorities. Toxicity to other organisms (algae, bacteria, earthworms, higher plants, mammals, and/or zooplankton) are also included, and the environmental hazard is evaluated from the toxicity to all of them considering the environmental distribution and the predicted doses or concentrations in the different environmental compartments (air, water, sediment, soil). In this case, the precision of single toxicity tests must be considered in relation to the precision of the other factors used in the hazard evaluation.

If, on the other hand, the objective is to determine toxicity to fish for derivation of water quality criteria, the major emphasis should be directed towards factors known or suspected to influence the toxicity of the chemical. In this case, variations in sensitivity due to abiotic factors like alkalinity, dissolved organics, hardness, pH, suspended solids, and temperature as well as biotic factors like species and nutritional status should be more important than variations due to precision.

Cases when precision and accuracy are of greater importance are those where small variations in toxicity of complex mixtures (e.g., effluents) are to be determined periodically (monitoring) or for comparative purposes (e.g., toxicity reduction evaluations, toxicity identification, legal testing for permits to discharge, etc.).

Determining the Accuracy and Precision of Test Methods

Variables known to influence toxicity to fish include alkalinity, dissolved oxygen, hardness, pH, temperature, species, developmental stage, and health (including nutritional status).[34,35] The objective of standardization is to minimize the influence of such variables. Furthermore, there is always a variation in sensitivity among individuals within a population. Therefore, toxicity values are usually accompanied by a statistical figure for the precision of the estimated value, the 95% confidence limits of the EC or LC_{50}. In addition to this, all measurements, those of toxicity included, have errors associated to their values. The quality of a measured value is reflected by its precision and its accuracy. Possible causes of variation in any determination are protocols, personal, experience, equipment, and laboratory environment. For toxicity tests, temporal variations in the sensitivity of test organisms, and the type of toxicant tested, will also contribute to this variation. All measurements of toxicity are empirical. Therefore, the accuracy of toxicity values are not determined in intercalibration exercises, only their precision.

Independent of the cause(s) of variation, the precision of any determination can be quantified operationally in three terms. These are:

1. **Replicability (r_o)** describing the variation between replicates (duplicates, triplicates, . . . , or parallels) during a single measurement (test). This variation is also called spatial variability.
2. **Repeatability (r)** describing the variation between repeated tests of the same chemical at the same laboratory, often made by the same person using the same equipment. This variation is also called intra-laboratory variability.
3. **Reproducibility (R)** describing the variation between tests of the same chemical using the same protocol but made by different persons at different laboratories and preferably at different times. However, for practical reasons determinations of reproducibility by intercalibrations are often made during a limited period of time. For the use of toxicity data, the desired interpretation of reproducibility is that any result obtained by the specific method has a precision that is known and is within the limits accepted by the scientific community.

For several reasons the variability will increase, when going from r_o to r to R. Most determinations made at established laboratories have low variabilities for r_o and r, but only for standardized methods do we know the variability for R. Any laboratory can determine and document the variability for r_o and r, but in order to determine the variability for R an intercalibration must be made.[24,36] Intercalibrations are primarily concerned with the variability for R, but they should preferably be designed to determine also the variability for r_o and r. The practical way to increase precision is to carry out parallel and preferably replicated tests on the samples. In parallel tests, r_o determines the minimum significant difference that is detectable.

Usually the variability of r_o, r, and R are expressed as CV, the coefficient of variation which is sometimes called RSD (relative standard deviation). CV is usually expressed as a percentage (SD/mean value $\times$ 100). Since a logarithmic series of concentrations is used in most toxicity tests, then the concentrations expressing toxicity (such as EC_{50}, LC_{50}, LOEC [Lowest Observed Effect Concentration], NOEC [No Observed Effect Concentration], and MATC [Maximum Acceptable Toxicant Concentration]) should be log-transformed prior to calculations of CV.[4]

PRECISION OF EMBRYO-LARVAL TOXICITY TESTS WITH FISH

Only those test methods that have been used at several laboratories and which have been subjected to at least one intercalibration are included in this evaluation. Unfortunately, in some cases the results from the intercalibrations have not been assessed separately for r_o, r, and R by the authors. However, since the primary results from different replicates and repeated tests have been given in the reports, calculations of CV for these three precision estimates are possible even if values for r_o, r, and R had to be estimated from separate sets of data. Therefore, response variability has been estimated only for controls, for responses such as fertilization rate, hatching time, growth, and survival.

However, since the variability may increase as a result of exposure to toxicants,[4] it should be emphasized that the variability of the final estimate of toxicity (LOEC, NOEC, MATC, ZEP = Zero Equivalent Point) is not the same as variability of response variables. Even if the inherent variability of response variables is important for precision, the variability in reaction to chemicals is of equal importance. Some responses may be more or less dependent upon specific chemical properties, and some chemicals will be more influenced by the period of exposure than others. Therefore, the estimated precision of a method is a complex function of variability not only of the method itself but also of the chemical(s) used to estimate it. Consequently, caution is needed when judging the precision of test methods found in intercalibrations, simply because chemicals chosen as toxicants in these exercises are not those which are likely to increase the variability of the data obtained. A high variability is expected for those chemicals that are unstable, volatile, adsorbed onto surfaces, have low aqueous solubility (high log $P_{octanol/water}$), and for which the chemical speciation is dependent upon water chemistry (e.g., weak acids and bases and heavy metals).

Sources of Information

So far only four EL tests have been assessed for reproducibility: (1) the embryo-larval test with the zebrafish (*Brachydanio rerio*)[3,4] and (2) the larval growth tests with the fathead minnow (*Pimephales promelas*),[25,31,37] (3) the sheepshead minnow (*Cyprinodon variegatus*),[38] and (4) the rainbow trout (*Onchorhyncus mykiss* formerly *Salmo gairdneri*).[30,39] A summary of the design of these tests is shown in Table 1.

Zebrafish EL Toxicity Tests

Three protocols have been produced for EL toxicity tests with the zebrafish (*Brachydanio rerio*) differing in duration and endpoints.[3,40,41] Two of the protocols[3,41] have been subjected to intercalibrations[4,41] and the results of these are summarized in Table 2 with emphasis on their precision of response variables.

Early Embryo Mortality and Fertilization Rate

The variable "early embryo mortality" in the controls can be used as an indicator of the biological quality of the starting material. Some

Table 1
Principal Design of Embryo-Larval and Larval Toxicity Tests with Fish

Fish species[a]	Developmental stage	Duration, exposure, vol./d	Response variables	Animals/ replicates	Endpoint(s)	Ref.
Fathead minnow	<24 h old larvae	7 d s-s[b]	survival (%) growth (7 d dw)	10-15/ 3-4	IC25 IC50	31 38
Sheepshead minnow		2.5 L/d			NOEC LOEC	
Fathead minnow	<36 h old embryos	7 d s-s	survival + terata (%)	10-15/ 3-4	EC1 EC5	31 38
Sheepshead minnow		2.5 L/d			EC10 EC50 NOEC LOEC	
Zebrafish	2-4 h old embryos	14 d s-s 0.1 L/d	median times for hatch and survival	30/ 1[d]	ZEP[c] NOEC LOEC	3
Rainbow trout	3-5 g larvae	28 d f-t 1200 L/d	growth rate (14 and 28 d body weight and length	16/ 1	NOEC LOEC	30 39

[a] Fish species used in tests are fathead minnow (*Pimephales promelas*), sheepshead minnow (*Cyprinodon variegatus*), zebrafish (*Brachydanio rerio*), and rainbow trout (*Oncorhynchus mykiss* previously called *Salmo gairdneri*).
[b] s-s (semi-static = static renewal), f-t (flow-through = dynamic).
[c] ZEP = Zero Equivalent Point = EC0.
[d] One replicate with 30 eggs (culled to 20 after 24 h) and 2 replicates for the control.

Table 2
Precision of Control Values for Variables Used in Embryo-Larval Toxicity Tests with the Zebrafish

Variable (unit)	Source of variation	Mean value	CV	N	Ref.
Early embryo mortality (%)	Replication r_O	28	16%	8	41
		23	37%	4	43
	Parental fish, r_{OP}	45	43%	4	43
	Laboratory, R	28	58%	8	41
		17	73%	5	4
Log median time for hatch (d)	Replication, r_O	0.47	8%	4	43
	Parental fish, r_{OP}	0.51	18%	4	43
	Repetition, r	0.45	10%	19	4
		0.41	17%	22	4
	Laboratory, R	0.45	11%	19	4
		0.41	17%	22	4
Log median survival time (d)	Replication, r_O	1.13	0.4%	4	43
	Parental fish, r_{OP}	1.13	1.7%	4	43
	Repetition, r	1.13	1.2%	19	4
		1.14	1.5%	22	4
	Laboratory, R	1.13	2.0%	19	4
		1.14	3.3%	22	4
Survival after 1 week (%)	Replication, r_O	97	0.8%	6	41
	Laboratory, R	95	8.5%	8	41
Survival after 6 weeks (%)	Replication, r_O	76	2.7%	6	41
	Laboratory, R	76	9.4%	8	41
Body length after 6 weeks (mm)	Replication, r_O	17.1	3.8%	6	41
	Laboratory, R	17.1	21%	8	41

authors prefer to call this variable "fertilization rate".[41,42] It is usually determined after 24 h (in warmwater fish like fathead minnows and zebrafish) when dead eggs (fertilized eggs with dead embryos and non-fertilized eggs) have turned white, while living embryos are transparent. Since the causes are unknown, both terms are misleading. In the absence of a proper term, I have called it "early embryo mortality". Some investigators prefer to eliminate or at least reduce this variable prior to the test by microscopic examination of the eggs before testing. Another possibility is to start the test with embryos aged for 24 h. Data from German[41,42] and Scandinavian laboratories[4,43] have shown that the early embryo mortality in the zebrafish is commonly in the range of 20 to 30%. In the early work with the zebrafish EL test at two Swedish laboratories (IVL, Stockholm and Department of Zoophysiology, Göteborg), we tried to reduce the early embryo mortality through more emphasis on careful broodstock management, and I still believe that such a reduction can be achieved (but not eliminated) by this means.

However, since early embryo mortality has not been found to be a sensitive variable for toxicants, I am not convinced that this is a very important issue. Larval nutrition in growth tests is probably more important, because growth is a response variable in some of these tests.

Hatching as an Indicator of Effects on Development

Natural embryo development in fish has been reviewed[44,45] and with emphasis on the effects of pollutants.[46,47] In contrast to many invertebrates, which have larval developmental stages with distinct morphological characteristics, the development in fish progresses gradually. Therefore, fish development is difficult to transform into discrete numbers. The only easily distinguishable developmental occurrence is the time of hatching. Hatching is also a process sensitive to environmental impact.[35] Even if its physiological control is not fully understood,[48,49] hatching has been found to respond to differences in pH[50] and to very low concentrations of zinc,[4] copper, and nickel.[43] However, its replicability is not as good as that for survival time,[4] and parental variability[43] should be considered in the evaluation of effects on hatching time. Fish embryos (zebrafish and rainbow trout) have also been proposed for use in teratogenicity screening.[51]

Larval Survival Time and Percentage Survival

Larval survival can be determined either in relation to time (in non-fed larvae)[3,4] or in relation to concentration (in fed larvae).[41] Survival time in non-fed zebrafish larvae is highly predictable with CVs for r_o, r, and R of 0.4, 1.2 to 1.5, and 2.0 to 3.3%, respectively. Percentage survival after 1 week had CVs for r_o and R of 0.8 and 8.5%, that is 2 to 4 times higher (Table 2).

Growth

Larval growth of zebrafish (Table 2), determined as body length after 6 weeks, had CVs of 3.8% for r_o and 21% for R. These values are about twice those for survival. Thus, even if growth may be a more sensitive process than survival, its variability is higher.

Larval Fathead Minnow Test

The 7-d larval test[29] was originally developed for mobile laboratory work. The first chronic life-cycle test with fish was made on fathead minnows[52] and it has been followed by numerous chronic as well as embryo-larval tests with the same species. Therefore, this test has the most extensive empirical support for extrapolating from effects in short-cuts (EL and larval tests are short-cuts to determine chronic toxicity) to the ultimate goal, the "no-effect" concentration for a (not any) fish species.[53] In addition, 25 years of experience on laboratory culture and testing with this species at the Environmental Research Laboratory in Duluth, and at other EPA, FWS, industry, university, and contract laboratories has probably brought the variability associated with this test close to its theoretical minimum. For 7 d control survival, the CV for R was actually lower than the CV for r_o (4.3 vs. 8 in Table 3). This is strange since the value for r_o was the mean from 23 replicated tests. One possible explanation for this illogical improvement when going from r_o to R could be the selection of results from those tests with a high control survival for publication; that is, repeating and not reporting tests with lower than 85% control survival.

Early embryo mortality is usually not reported in the fathead minnow EL or larval tests. Pickering[54] reported that during his extensive study about 15% of the eggs were unfertilized. Control survival from Day 1 to Day 8 in EL and larval tests was usually around 95%, that is, the same as in zebrafish tests,[41] and the reproducibility was similar. The CV for R of 4.3% in the fathead minnow (Table 3) may be too low for the reasons given above. Therefore, the CV for r_o of 8 might be better compared with the CV for R of 8.5% found for the zebrafish (Table 2).

Table 3
Precision of Control Values for Variables Used in Embryo-Larval Toxicity Tests with the Fathead Minnow

Variable (unit)	Source of variation	Mean value	CV	N	Ref.
7 d survival (%)	Replication, r_o	94	8%	23	54
	Repetition, r	94	6%	23	54
	Laboratory, R	97	4%	10	60
7 d body weight (mg dw)	Replication, r_o	0.67	9%	23	54
	Repetition, r	0.67	24%	23	54
	Laboratory, R	0.50	36%	10	54

Larval growth (dry body weight) in the 7 d test with fathead minnows had CVs for r_0, r, and R of 9, 24, and 36% (Table 3). For 6-week tests, CVs for body length in the zebrafish were 3.8% for r_0 and 21% for R. Thus, for survival and growth, the control variabilities are similar in the fathead minnow and the zebrafish tests.

Larval Growth Test with Rainbow Trout

The larval growth test with the rainbow trout proposed by Crossland[30] has been the subject of one intercalibration.[39] A major advantage of using rainbow trout is its large size (3 to 5 g) so that individual measurements can be made of body weight. Also its commercial use in fish culture has led to the development of synthetic diets (dry food) and automatic feeders. Another advantage of using rainbow trout is its high sensitivity to chemicals in general.[55] A practical disadvantage is the requirement of larger volumes of water (Table 1), making it impractical to use reconstituted water. In tests with effluents, the transport and preservation of large quantities of water samples (by cooling or freezing) is a serious obstacle.

Estimating the "No-Effect" Concentration

In acute toxicity tests, the effect (mortality or immobility) is used to derive the concentration causing an effect in 50% of the test organisms. A chronic EL or L subchronic toxicity test is more ambiguous because the objective is to determine the "no-effect" concentration during chronic exposure. For statistical reasons a "no-effect" concentration cannot be established positively. All statistical tests are based on rejection of a null-hypothesis. Therefore, the LOEC can be identified by a statistical test.[56–59] NOEC is the next lower concentration, and the "no-effect" concentration is assumed to be the geometric mean (logarithmic midpoint) of LOEC and NOEC (GM-LOEC-NOEC). Alternatively, the maximum "no-effect" concentration can be defined as that producing a predefined effect, e.g., an 'x'% decrease in survival, body weight or length, or other measured responses. Then this predefined effect level ('x') must be considered in relation to the variability of the response variable. In the evaluation of data, some statistical tests (e.g., Dunnett's test), besides identifying the LOEC, also determine the Minimum Significant Difference (MSD, P=0.05). For 20 7-d larval toxicity tests made with 4 toxicants in the fathead minnow using 4 replicates with 25 larvae, the MSD for percentage survival and growth (body dw) averaged 14 and 16% with CVs of 6 and 7%, respectively.[54]

In all toxicity tests the concentrations are selected in advance, and if a hypothesis testing procedure is used in the evaluation as in the NOEC-LOEC-approach, the selection of test concentrations and the dilution factor will always have a major influence on the estimated "no-effect" concentration. However, since most chronic EL and L toxicity tests have been evaluated by hypothesis testing, the GM-LOEC-NOEC has been used to determine their replicability.[4,54] In other papers[25,37,60] the 7-d IC_{50} (Inhibitory Concentration 50%) or LC_{50}s have been used and the reproducibility has been expressed as the CV for arithmetic values.

Hypothesis Testing vs. Regression of the Dose-Response Relationship

In many intercalibrations, the participating laboratories are not asked for their determination of the "no-effect" concentration. They only report primary data, often from predefined concentrations of reference toxicants. The statistical evaluation is then made uniformly by the executive officer. In the zebrafish ring test using two toxicants, Cr and Zn,[4] where both a hypothesis testing and a manual graphical regression method (log median effective time vs. log concentration on paper and eye-fitting) was used, the "no-effect" concentrations were estimated independently by two persons, the supervisor at the participating laboratory (E1) and the executive officer for the ring test (E2). The estimated "no-effect" concentration derived by graphical method is called ZEP, which is the lowest concentration with a measured response equal to the control(s), i.e., EC_0. The difference in "no-effect" concentrations for effects on survival time and precision were similar for both "no-effect" concentration estimations with CVs for r of 5 to 11% and CVs for R of 8 to 24% (Table 4). However, for the effect of Zn on hatching time, which was more sensitive than that for survival time, the CVs were much higher (100 to 200%). Furthermore, estimates of ZEP for effects on hatching were more subjective. One of the two estimators (E2) achieved ZEP to be only half that of the other (E1).

Table 4
Precision for Determination of "No Effect" Concentrations in Embryo-Larval Toxicity Tests with the Zebrafish[4]

Toxicant	Estimated concentration	Estimator	Log mean	CV_r (%)	CV_R (%)	N
$K_2Cr_2O_7$	GM-NOEC-LOEC[a]	E1	1.66	6.5	24	9
	for survival	E2	1.74	7.2	19	8
	ZEP[b] for survival	E1	1.66	4.7	23	9
		E2	1.52	11.1	24	9
$ZnSO_4x7H_2O$	GM-NOEC-LOEC	E1	1.55	8.4	14	10
	for survival	E2	1.52	10.3	15	10
	ZEP for survival	E1	1.52	9.7	11	10
		E2	1.50	8.1	10	10
	GM-NOEC-LOEC	E1	0.61	80	96	9
	for hatching	E2	0.60	76	107	6
	ZEP for hatching	E1	0.58	102	102	10
		E2	0.30	193	193	9

[a] Geometric Mean of NOEC and LOEC.
[b] Zero Equivalent Point.

Subjective Error of "No-Effect" Concentration Estimates

Table 4 shows that for nine tests with chromium, E1 and E2 GM-NOEC-LOEC estimates for survival were identical in 6 out of 9 tests and in 2 out of 9 tests they differed by a factor 2 (reciprocal of the dilution factor of 0.5). In 1 of the 9 tests, one of the estimates (E2) was below the lowest test concentration. For 10 tests with zinc, E1 and E2 GM-NOEC-LOEC concentrations for survival were identical in 9 tests and different by a factor 2 in 1 test. Thus, subjectively introduced variability is less than that associated by reproducing the tests at different laboratories, except for the effects of Zn on hatching, as mentioned above.

DISCUSSION

The areas addressed below are the reproducibility of the standardized methods in general, the existence of different endpoints, experimental design with emphasis on dilution factor, and the need for and use of replicates.

Table 5
Replicability, Repeatability, and Reproducibility for Determination of 7-d LC50 and 7-d IC50 in Larval Tests with Fathead Minnows[37]

Toxicant	Test period	Endpoint	Mean value	CV_{rO}	CV_r	CV_R
PCP	1	LC50	277 ug/L	15.6	23.3	43.7
	2		285 ug/L	10.9		42.8
	1	IC50	296 ug/L	16.4	22.7	32.4
	2		289 ug/L	9.2		37.6
$K_2Cr_2O_7$	1	LC50	19 mg/L	10.0	14.5	24.1
	2		20 mg/L	6.9		27.9
	1	IC50	8 mg/L	16.2	17.6	27.5
	2		10 mg/L	26.1		88.0
RE 301[a]		LC50	1.0%	9.1		31.3
		IC50	0.5%	9.1		40.0
RE 401[b]		LC50	38%	7.9		25.6
		IC50	33%	13.0		22.4
UWS 501[c]		LC50	3.7%	5.9		37.5
		IC50	2.1%	16.1		61.9

[a] Refinery effluent 301.
[b] Refinery effluent 401.
[c] Utility Waste Stream 501.

Is the Reproducibility of Embryo-Larval Toxicity Tests Satisfactory?

The efforts to standardize the embryo-larval test methods for fish have been very productive. In general, all endpoints (LOEC, NOEC, IC_{25}, EC_{10}, ZEP) used in this test have a reproducibility of about $\leq \pm 2$ x the median endpoint with effects on survival having a somewhat better inter-laboratory precision than effects on growth (Table 5). In a recent evaluation[25] of the interlaboratory precision obtained in the intercalibration of the 7-d larval survival and growth test with 5 toxicants (7 cases[37] summarized in Table 6), it was found that NOEC values for survival that were $> \pm 2$ concentrations from the median NOEC were found in 6 to 18% (mean ± SD: 11 ± 5%) of the tests with 2 replicates, and in only 0 to 14% (4 ± 6%) of the tests with 4 replicates. In the case of NOEC values for growth, these figures were 0 to 28% (16 ± 15%) with 2 replicates and 0 to 33% (11 ± 11%) with 4 replicates. Thus, in

Table 6
Summary of Repeatability (r) and Reproducibility (R) Obtained in Intercalibrations of Early Life Stage Toxicity Tests with Fish

Species[a]	Medium[b]	Toxicant[c]	Variables[d]	Endpoint[e]	Mean value (mg/L)	CV_r (%)[f]	CV_R (%)[g]	N[h]	Ref.[i]
C.v.	asw	Cu	s or g	IC25	0.07	42		8	25
	nsw				300	33		5	25
	asw	SDS			1.5	31		6	25
	nsw				2.2	28		4	25
	sw	ind. effl.			5.5%		44	7	25
	asw	Cu	s and t	EC10	202	3		4	25
	asw	SDS			1.9	35		4	25
M.b.	nsw	Cu	s or g	IC25	210	44		5	25
	nsw	SDS			1.3	43		5	25
P.p.	fw	diquat	s and t	LC1	1.5	41		5	25
		$CdCl_2$		LC1	0.0068	62		5	25
	5 fw's	Cr^{+6}	s or g	IC25	5.1	27		10	25
	rfw	PCP	s	LC50	0.281	23	43	20	37
			g	IC50	0.293	23	35	20	37
		$K_2Cr_2O_7$	s	LC50	20	15	26	20	37
			g	IC50	9	18	58	20	37
		RE 301	s	LC50	1%		31	10	37
			g	IC50	0.5%		40	10	37
		RE 401	s	LC50	38%		26	10	37
			g	IC50	33%		22	10	37
		UWS 501	s	LC50	3.7%		38	10	37
			g	IC50	2.1%		62	10	37
		Cr^{+6}	s or g	IC50	6.9		31	9	60
B.r.	afw		st	ZEP	18	33	52	9	4
		Zn^{+2}			7.7	26	35	10	4
			ht		1.2	125	125	9	4
O.m.	fw	DCA	$SGr_{0\text{-}28d}$	LOEC	0.341		58	10	39
		LAS			1.900		19	3	39
		M-par.			0.318		34	4	39

[a] Species abbreviations are by first letter for genus and species for *Cyprinodon variegatus* sheepshead minnow), *Menidia beryllina* (inland silverside), *Pimephales promelas* (fathead minnow), *Brachydanio rerio* (zebrafish), and *Oncorhynchus mykiss* (rainbow trout).

[b] Abbreviations used for test media are: a = artificial (reconstituted), n = natural, sw = sea water, fw = freshwater. 5 fw's means that 5 different kinds of freshwater were used in this comparison.

[c] Abbreviations for toxicants used are: Cu = copper, SDS = sodium dodecyl sulfate, ind. eff. = industrial effluent, PCP = sodium pentachlorophenate, RE 301 = refinery effluent 301, RE 401 = refinery effluent 401, UWS 501 = utility waste 501, DCA = 3,4-dichloroaniline, LAS = linear alkylbenzene sulphonate, and M-par. = methyl parathion.

[d] Variable abbreviations are: s = survival, g = growth (7 d body weight), t = teratogenicity, st = survival time, ht = hatching time, $SGR_{0\text{-}28d}$ = Standard Growth Rate between day 0 and 28.

[e] Endpoints are abbreviated as follows: IC25 = Inhibitory concentration 25%, EC10 = Effective concentration 10%, LC1 = Lethal concentration 1%, ZEP = Zero Equivalent Point (= EC0), LOEC = Lowest Observed Effect Concentration. LOEC, which is not suitable for statistical evaluation, was used in the absence of other given endpoints.

[f] CV_r = Coefficient of Variation in repeated tests (intra-laboratory precision).

[g] CV_R = Coefficient of variation for test made at different laboratories (inter-laboratory precision).

[h] N = number of data (tests) used for determination of mean and CV.

[i] Reference to source of information for either reported CVs or raw data used for calculation of mean value and CV for r or R.

nine out of ten tests, an estimated value for NOEC is expected to be within a factor of 2 of the median value obtained in a multi-laboratory test.

Survival vs. Growth as Response Variables

A major reason for higher CVs for growth than for survival is probably related to larval nutrition. For larvae of small species (e.g., fathead minnow and zebrafish) no appropriate synthetic reference diets are presently available. Therefore, live food (*Brachionus, Artemia*) are the commonly used diets. Their nutritional value as well as the exact amounts given to the fish may vary from one test to another (increasing CV for r) and from one laboratory to another (increasing CV for R). For feeding fathead minnow larvae, the U.S. EPA recommended the use of *Artemia* from a certified source, with frequent feeding (3 times per day providing an *ad lib* regime of live food during the 16 h light period), as an appropriate diet.[31] For rainbow trout growth tests, various methodological problems dealing with growth determinations were studied by Crossland.[30] However, it must also be realized that growth is a very complicated physiological process[62] which is subjected to individual rhythms.[63] Therefore, its interpretation in an ecological perspective is more difficult than that for survival.

The Use of Different Endpoints in EL Toxicity Tests is Confusing

The use of multiple endpoints can be confusing for the users of toxicity information, who are not always familiar with the details relating to the methodology. Therefore, it is the responsibility of the aquatic toxicologists to make a decision on the most relevant endpoints. Even the use of two endpoints based on different response variables (survival and growth) can give rise to endless debates on the relative sensitivity of these variables, their ecological relevance, the statistical vs. biological significance of findings, etc. Fortunately, these variables often provide the same "no-effect" concentration, but considerable difficulties arise when there are differences. It may even happen that one variable is more sensitive in the first test run and the other is more sensitive in the second test run with the same chemical using the same test.[25]

End of display - Press <RETURN> to continue, P for <Previous>, PR to Print >Q

PRG1 Northern Arizona U. Charge Receipt 11/02/1994 13:42:04

TILLMAN MARIAH D010448894

1 A10207706048 Progress in standardization of aqu 11/23/94

However, this is a minor problem compared to the use of a multitude of methods for evaluation of the same data. Often they provide very similar results, but again when they do not there are difficulties. This problem is exaggerated with the NOEC-LOEC approach, where the difference in endpoint may be 2 to 3 times apart, depending upon the dilution factor (0.5 vs. 0.3). Thus, the difference in endpoint due to statistical treatment may be as great as, for example, the reduction of toxicity achieved between commonly applied treatment methods for effluents such as sewage treatment and aerated lagoons, where parallel toxicity tests must be used to obtain a proper comparative evaluation.

The philosophy of using the EC_{50} as an endpoint in toxicity tests is its higher precision compared to other endpoints for a fixed sample size.[61] This is because at this concentration a defined change in concentration will always cause a maximum change in response. Therefore, the EC_{50} is the most precise estimate of toxicity, and theoretically the EC_{50} should have the smallest CVs for r_o, r, and R. Values in Table 5 are based on such endpoints.

However, the ratio between the EC_{50} and the "no-effect" concentration will differ among chemicals because it depends on the slope of the dose response relationship. Chemicals with a low slope will have a high EC_{50}-EC_0-ratio. Such chemicals will also have wider confidence limits, both for the EC_{50} and, even more so, for the "no-effect" concentration. Refinements of test designs and statistical tests can improve the capability to determine the "no-effect" concentration, but these refinements can not change the inverse chemical-specific relationships between slope of the dose-response curve and the confidence limits for the effect concentration.

The intuitive interpretation of the expression "no-effect" concentration is "zero-effect" concentration or EC_0. This is the interpretation of "no-effect" of open-minded students. However, this interpretation highlights the problem to defining the term "biological zero". A practical solution to this is to confine the "effect" to the context of the test, and to use the control response as the "zero response". If effects are determined over the entire dose-response relationship (from 0 to 100% effect) then the ZEP can be determined, after appropriate transformations, as the intersection between the fitted dose-response relationship and the zero response as shown in Figure 1. The precision of ZEP is equal or better than for the geometric mean of NOEC and LOEC.[4, 43]

For obvious reasons, a toxicity test showing the complete dose-response relationship is more informative than one showing only the

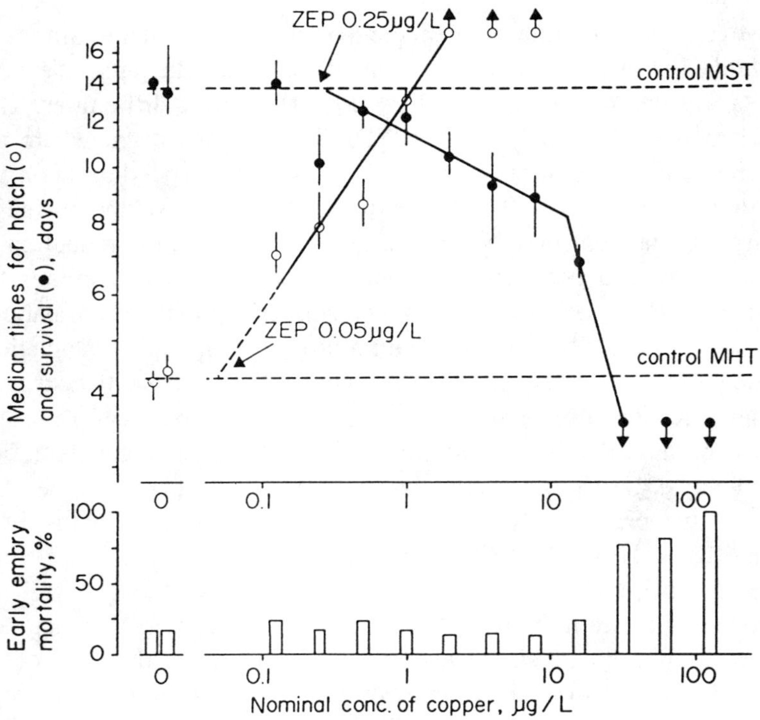

FIGURE 1. Example of dose-response relationship in an embryo-larval toxicity test with copper in the zebrafish. In high concentrations (> 10 µg/L), even early embryos were killed and in intermediate concentrations (1 – 10 µg/L), embryos were killed prior to hatching. In lower concentrations, survival time was reduced and hatching time was delayed. Zero Equivalent Points (ZEPs) for effects on survival (0.25 µg/L) and hatching (0.05 µg/L) were determined graphically as shown in the figure. Corresponding geometric means for NOEC and LOEC were 1.41 µg/L and <0.125 µg/L, respectively.[43]

portion where the minimum detectable toxicity is present. Furthermore, a test design directed exclusively towards the lower portion of the dose-response curve must be capable of detecting a smaller effect and, therefore, the experiment should have more observations (animals, replicates) in this region. In the hypothesis testing design, the maximum emphasis should be directed towards three treatments: (1) the control, (2) the assumed NOEC, and (3) the assumed LOEC. The precision of the "no-effect" concentration (GM-NOEC-LOEC) is then determined by the proximity of the NOEC to the LOEC (= the dilution factor) and the statistical power of the experiment (df = N-1).

Why Not Fix the Dilution Factor?

A dilution factor of 0.5 was used in the intercalibrations of the fathead minnow 7-d test[37] and the zebrafish 14-d test.[4] Smaller dilution factors (wider spacing of concentrations) were used in the intercalibrations of the 28-d rainbow trout growth test[39] and the 6-week zebrafish test.[41] A dilution factor of 0.5 is a good choice for many reasons. It is related to the precision of the "no-effect" concentration determined in EL toxicity tests. It provides optimum precision in diluting test solutions (toxicant solution and diluent are used in equal quantities meaning equal precision), and with the usual number of test organisms per concentration (20 to 40) at least two partial effect concentrations are usually obtained. Two partial effects are a requirement for determinations of confidence limits by probit analysis or moving average methodology.[61]

Does Replication Increase the Precision and Accuracy of EL Toxicity Data?

Replicates are often used in toxicity tests, but rationales for using them are rarely discussed. Some of the true and false arguments in favor of replicates are listed in Table 7.

In toxicity tests, replicates are made to separate artifacts (manipulative-induced effects) from true (toxicant-induced) effects. In the laboratory, the influence of manipulative effects can be minimized by experimental design though randomization to treatment and circulation of positions, observations, and manipulative sequences.

In EL toxicity tests, the variation in response between replicates is usually small and the CVs for r_0 shown in Tables 2, 3, and 5 reflects this. Furthermore, in relation to r and R the CVs increase by about a factor of 2 for each step (r_0 to r to R). Therefore, in the screening of chemicals and products, instead of using 4 replicates at 1 laboratory, one should use 1 replicate at 4 laboratories for the same test with the same chemical. Such an approach would give a more realistic figure for the expected species-specific variability in response to the toxicant under study. Furthermore, if the same measurement has been made at different laboratories, outliers can be identified and a quality control for accuracy is then made possible. This is not possible when using results from replicates made at only one laboratory. Thus, in the absence of "true" values for toxicity, the CV for r_0 determines the

Table 7
True and False Arguments in Favor of Using Replicates in EL Toxicity Tests

Argument	True/false	Comment
Visibility	True	Small dead organisms may disintegrate rapidly, especially in waste waters. Therefore survival counts are preferable to mortality counts. Moving objects like swimming larvae are difficult to count. For small rapidly moving objects counted under the microscope, counting error increases with more than 7 objects.
Randomization improves	True/false	The effect of poor randomization is counteracted by proper replication. However, with a proper randomization, the expected improvement is minimal.
Spatial interaction is counteracted	True/false	Spatial effects of temperature, light, humidity, and the time sequence of manipulations during renewals of test solutions, feeding, recordings, etc. are counteracted by circulation and randomization. These measures taken may be as important as replication.
Interaction between test organisms	True	Test organisms may interact by aggression, cannibalism, competition, and transmission of pathogens. These events may be part of the effect (primary or secondary) of the toxicant or occur at random. The use of replicates may help to distinguish between these causes.
Human error (mistakes)	True/false	Mistakes are made also in toxicity tests. The impact of a single mistake is inversely related to the number of replicates, i.e., the impact of a single mistake affecting only one of several replicates is less disastrous than that on a single treatment. However, the probability of making a mistake increases with the number of operations made, i.e., with the number of replicates.
Replication increases precision	True/false	Measured values from replicates which are very close (which they should be in a properly designed experiment) does not increase the precision of the estimated value. This is because values from either one of the replicates are then close to the mean for the replicates.
Replicate values for growth and survival have different meanings	True	Survival is a discontinuous (discrete) variable, the precision of which is determined by the number of organisms tested (N). Growth, being a continuous variable, is determined either on individuals or on groups of individuals (pooled mean value). Pooling reduces the variance but it also reduces the sample size (N).

precision of the experimental value while the CV for R determines the accuracy of the toxicity value. The extra cost of this multi-laboratory approach to chemical testing can be compensated for in several ways.

One is to use only two independent laboratories in the first run. If the results from the two laboratories are within the expected precision (ratio ≤ 2) then they can be considered valid, but if not (ratio > 2) then a test at one or two additional laboratories must be made. Thus, a general saving in testing effort can pay for the extra effort required to solve doubtful cases. The expected frequency of such cases is 10 to 20% based on the intercalibration figures from the fathead minnow 7-d test[37,65] as discussed above.

RECOMMENDATIONS

1. Short-term tests with embryos and larvae in common use have a repeatability and reproducibility equal to that of acute toxicity tests. Therefore, they can be recommended to use for regulatory purposes.
2. Means to increase the precision of these tests in monitoring of effluents and toxicity reduction evaluations should be directed towards the use of parallel, replicated tests of proper design and the use of endpoints with optimal precision. These are EC_{50}s for mortality (LC_{50}), growth (IC_{50}), or other suitable variables.
3. The use of multiple-replicated tests for regulatory control of chemicals and products will not increase precision per se. Instead, they can give a false impression of accuracy. Several tests conducted at different laboratories with fewer replicates will provide a more accurate estimate of inherent chemical toxicity for a chemical than one test with several replicates conducted at one laboratory.
4. Embryo-larval tests should preferably be designed to determine toxicity over the entire dose-response relationship (from EC_0 to EC_{100}). Such data can serve a multitude of purposes. Those directed only towards the lower portion often fail to provide any information on inherent toxicity.
5. The potential to estimate chronic toxicity for different species of fish has not been explored to any extent.[32,33] Tests with different species could provide an empirical alternative to the application factor approach for chronic toxicity. Such studies

should be linked to the toxic action of different chemicals because these are important in the development of Quantitative Structure Activity Relationships (QSARs).

6. Studies on the effect of parental exposure[64] is another area that could be explored. This effect may possibly be used as an index of reproductive effects in wild fish captured at various locations.

7. In the intercalibrations of EL toxicity tests made so far[4,37,39,41] no indication of intra-specific variability among tested stocks have been detected. However, a slight parental influence has been detected[43] which emphasizes the need for proper randomized designs when conducting EL toxicity tests.

ACKNOWLEDGMENTS

The results summarized in this review have been produced by the careful and dedicated work of too many people to mention here, working at government, university, industry, and consulting laboratories. Special funding for intercalibration activities is often lacking. Therefore, I would like to thank especially those who have offered time and reallocation of resources needed for these activities. Furthermore, I would like to thank those engaged in standardization of methodology at all levels and who are rarely mentioned in any scientific publication.

REFERENCES

1. Penny, C. and Adams, C. 1863. Fourth report, Royal Commission on Pollution of rivers in Scotland. Vol. 2. Evidence. London. (cf. Hunn, 1989[12]).
2. Powers, E.B. 1917. The goldfish (*Carassius*) as a test animal in the study of toxicity. III. *Biol. Monogr.* 4:7–73.
3. SIS (Standardiseringskommissionen i Sverige). 1988. Vatenundersökningar — Bestämning av toxicitet för embryoner och yngel av sötvattenfisk-Semistatisk metod. Svensk standard SS 02 81 93, 12 pp. (Swedish standard SS 02 81 93, English translation available upon request to the author).

4. Dave, G., Damgaard, B., Grande, M., Martelin, J.E., Rosander, B. and Viktor, T. 1987. Ring test of an embryo-larval toxicity test with zebrafish (*Brachydanio rerio*) using chromium and zinc as toxicants. *Environ. Toxicol. Chem.* 6:61–77.
5. Belding, D.L. 1927. Toxicity experiments with fish in reference to trade waste pollution. *Trans. Am. Fish. Soc.* 57:100–119.
6. Hunn, J.B., and Allen, J.L. 1974. Movement of drugs across the gills of fishes. *Annu. Rev. Pharmacol.* 14:47–55.
7. Berger, E. 1930. Untershiedliche Wirkungen gleicher Ionen und lonengemische auf verschiedene Tierarten. (Ein Beitrag zur Lehre vom Ionenantagonismus.) *Pflugers Archiv. Physiol.* 223:1–39.
8. Fluckiger, E. und Fluck, H. 1949. Ein Kunstliches Milieu fur das Zeuchten von Daphnien im Laboratorium. *Experientia*, 5(12):485.
9. Freeman, L. and Fowler, I. 1953. Toxicity of combinations of certain inorganic components to *Daphnia magna* Straus. *Sewage Ind. Wastes,* 25:1191–1195.
10. Doudoroff, P. and Katz, M. 1950. Critical review of literature on the toxicity of industrial wastes and their components to fish. I. Alkalies, acids, and inorganic gasses. *Sewage Ind. Wastes,* 22:1432–1458.
11. Doudoroff, P. and Katz, M. 1953. Critical review of literature on the toxicity of industrial wastes and their components to fish. II. The metals, as salts. *Sewage Ind. Wastes,* 25:802–839.
12. Hunn, J. 1989. History of acute toxicity tests with fish, 1863–1987. U.S. Fish Wildl. Serv., Invest. Fish Control, 98. 10 pp.
13. American Public Health Association, American Water Works Association, and Water Pollution Control Federation. 1960. Standard methods for the examination of water and wastewater. 11th ed. American Public Health Association, New York. 626 pp.
14. Committee on Methods for Toxicity Tests with Aquatic Organisms. 1975. Methods for acute toxicity tests with fish, macroinvertebrates, and amphibians. U.S. Environmental Protection Agency, Ecological Research Series, EPA-660/3-75-009. 61 pp.
15. Marking, L.L. 1975. Effects of pH on toxicity of antimycin to fish. *J. Fish. Res. Board Can.* 32:769–773.
16. American Society for Testing and Materials. 1980. Standard practice for conducting acute toxicity tests with fishes, macroinvertebrates, and amphibians. E729–80. American Society for Testing and Materials, Philadelphia, Pa. 25 pp.
17. American Public Health Association, American Water Works Association, and Water Pollution Control Federation. 1981. Standard Methods for examination of water and wastewater. 15th ed. American Public Health Association, Washington, D.C. 1134 pp.

18. Reiff, B., Lloyd, R., How, M., Brown, D. and Alabaster, J.S. 1979. The acute toxicity of eleven detergents to fish: results of an inter-laboratory exercise. *Water Res.* 13:207–210.
19. European Inland Fisheries Advisory Commission. 1975. Report on fish toxicity testing procedures. FAO/EIFAC, Rome, Tech. Pater 25, 25 pp.
20. International Organization for Standardization, 1976. Summary of results of acute toxicity tests with zebra fish carried out according to the procedure proposed by ISO/TC147/SC5/WG3 (Secretariat -6) 18, Nov. 11, 1976.
21. Commission of the European Communities. 1979. Inter-laboratory ring test concerning the ecotoxicity of a chemical substance with respect to fish. Institut National de Recherche Chimique Appliquee. Centre de Recherche, 917 10 Vert-le-Petit, France. 18 pp.
22. Wellens, H. 1982. Vergleich der Empfindlichkeit von *Brachydanio rerio* und *Leuciscus idus* bei der Untersuchung der Fischtoxizität von chemischen Verbindungen und Abwässern. *Z. Wasser Abwasser-Forsch. (Ger.)* 15:49–52.
23. International Organization for Standardization, 1984. Water quality - Determination of the acute lethal toxicity of substances to a freshwater fish [*Brachydanio rerio* Hamilton-buchanan (Teleostei, Cyprinidae)]. Part 1:static method, ISO 7346/2–1984(E), Part 3: Flow-through method, ISO 7346/3-1984(E).
24. How, M.J. 1980. The application and conduct of ring tests in aquatic toxicology. *Water Res.* 14:293–296.
25. U.S. Environmental Protection Agency. 1991. Technical support document for water quality-based toxics control. EPA/505/2-90-001.
26. Macek, K.J. and Sleight, B.H. 1977. Utility of toxicity tests with embryos and fry of fish in evaluating hazards associated with the chronic toxicity of chemicals to fishes. In: F.L. Mayer and J.L. Hamelink, Eds., Aquatic Toxicology and Hazard Evaluation, ASTM STP 634, American Society for Testing and Materials, Philadelphia, Penn. pp. 137–146.
27. McKim, J.M. 1977. Evaluation of tests with early life stages of fish for predicting long-term toxicity. *J. Fish. Res. Bd. Can.* 34:1148–1154.
28. McKim, J.M. 1985. Early life stage toxicity tests. In: G.M. Rand and S.R. Petrocelli, Eds., Fundamentals of Aquatic Toxicology, Hemisphere Publ., Washington, D.C., pp. 58–95.
29. Norberg, T.J. and Mount, D.I. 1985. A new fathead minnow (*Pimephales promelas*) subchronic toxicity test. *Environ. Toxicol. Chem.* 4:711–718.
30. Crossland, N.O. 1985. A method to evaluate effects of toxic chemicals on fish growth. *Chemospere,* 14:1855–1870.

31. U.S. Environmental Protection Agency. 1989. Short-term methods for estimating the chronic toxicity of effluents and receiving waters to freshwater organisms. 2nd ed. EPA 600/4-89/001 and Suplement EPA 600/4-89/001A. Cincinnati, OH.

32. McKim, J.M., Eaton, J.J., and Holcombe, G.W. 1978. Metal toxicity to embryos and larvae of eight species of freshwater fish-II. Copper. *Bull. Environ. Contam. Toxicol.* 19:608–616.

33. Sauter, S., Buxton, K.S., Macek, K.J., and Petrocelli, S.R. 1976. Effects of exposure to heavy metals on selected freshwater fish. Toxicity to copper, cadmium, chromium and lead to eggs and fry of seven fish species. U.S. Environmental Protection Agency, EPA-600/3-76-105, Cincinnati, OH. 75 pp.

34. Sprague, J.B. 1985. Factors that modify toxicity. In: G.M. Rand and S.R. Petrocelli, Eds., Fundamentals of Aquatic Toxicology, Hemisphere Publ., Washington, D.C., pp. 124–163.

35. Dave, G. 1986. Toxicity testing procedures. In: S. Nilson and S. Holmgren, Eds., *Fish Physiology: Recent Advances.* Croom Helm. pp. 170–195.

36. Waibel, J. 1978. Plaunung und Auswertung von Ringversuchen zur Ermittlung der wiederholbarkeit und Vergleichbarkeit. *Fette, Seifen, Anstrichmittel,* 80:410–413.

37. DeGraeve, G.M., Cooney, J.P., Pollock, T.L., Reichenbach, N.G., Dean, J.H., Marcus, M.D., and McIntyre, D.O. 1988. Intra- and interlaboratory study to determine the reproducibility of the seven-day Fathead minnow larval survival and growth test. API Publ. No. 4468, American Petroleum Institute, Washington, D.C. 220 pp. (also in Reference 65).

38. U.S. Environmental Protection Agency. 1988. Short-term methods for estimating the chronic toxicity of effluents and receiving waters to marine and estuarine organisms. EPA 600/4-87/028. Cincinnati, OH.

39. Ashley, S. and Mallet, M.J. 1990. EEC ring test of a method for determining the effects of chemicals on the growth rate of fish. Final report to the Commission of the European Communities. CEC 2600-M. 76 pp. + appendices.

40. Kristensen, P., Balk, F., Damgaard, B., Tyle, H. and Van Leeuven, C.J. 1988. Short-term early life stage test (short-term ELS test) on zebra fish (*Brachyodanio rerio*, Hamilton-Buchanan, *Teleostei, Cyprinidae*). Methods for determination of ecotoxicity (EEC). Water Quality Institute, Denmark and AKZO Corporate Research Laboratory, The Netherlands.

41. Nagel, R., Bresch, H., Caspers, N., Hansen, P.D., Markert, M., Munk, R., Scholz, N., and ten Höfte, B.B. 1991. Effect of 3,4-dichloroaniline on the early life stages of the zebrafish (*Brachydanio rerio*): results of a comparative laboratory study. *Ecotox. Environ. Safety,* 21:157–164.

42. Bresch, H., Gode, P., Hamburger, B., Hansen, P.D., Junke, I., Krebs, F., Lillelund, K., Markert, M., Munk, R., Nagel, R., Nusch, E.A., Scheubel, J., and Spieser, O.H. 1986. Deliberations and investigations of the DIN working group "fish test" on a long term test according to the German Chemicals Act. *Z. Wasser-Ab-wasser-Forsch.* 19:47–51.

43. Dave, G. and Xiu, R. 1991. Toxicity of mercury, cooper, nickel, lead, and cobalt to embryos and larvae of zebrafish, *Brachydanio rerio. Arch. Environ. Contam. Toxicol.* 21:126–134.

44. Balon, E.K. 1985. Early life history of fishes: New developmental, ecological and evolutionary perspectives. Junk Publ., Dordrecht. 280 pp.

45. Blaxter, J.H.S. 1988. Pattern and variety in development. In: W.S. Hoar and D.J. Randall, Eds., *Fish Physiology*, Vol. XIA, Academic Press, pp. 1–58.

46. Rosenthal, H. and Alderdice, D.F. 1976. Sublethal effects of environmental stressors, natural and pollutional, on marine fish eggs and larvae. *J. Fish. Res. Board Can.* 33:2047–2065.

47. von Westerhagen, H. 1988. Sublethal effects of pollutants on fish eggs and larvae. In: W.S. Hoar and D.J. Randall, Eds., *Fish Physiology,* Vol. XIA, Academic Press, pp. 253–346.

48. Denuce, J.M. 1985. How embryos escape from their envelopes: A new look at the (phylogenetically) old problem of hatching. *Meded. Kon. Acad. Wet. Belgi.* 46:1–30.

49. Yamagani, K. 1988. Mechanisms of hatching in fish. In: W.S. Hoar and D.J. Randall, Eds., *Fish Physiology*, Vol. XIA, Academic Press, pp. 447–499.

50. Dave, G. 1985. The influence of pH on the toxicity of aluminum, cadmium and iron to eggs and larvae of zebrafish (*Brachydanio rerio*). *Ecotox. Environ. Safety,* 10:253–267.

51. Van Leeuwen, C.J., Grootelaar, E.M.M., and Niebeek, G. 1990. Fish as teratogenicity screens: a comparison of embryotoxicity between fish and birds. *Ecotox. Environ. Safety,* 20:42–52.

52. Mount, D.I. and Stephan, C.E. 1967. A method for establishing acceptable toxicant limits for fish — malathion and butoxyethanol ester of 2,4-D. *Trans. Am. Fish. Soc.* 96:185–193.

53. Norberg-King, T.J. 1989. An evaluation of the fathead minnow seven-day subchronic test for estimating chronic toxicity. *Environ. Toxicol. Chem.* 8:1075–1089.

54. Pickering, Q.H. 1988. Evaluation and comparison of two short-term fathead minnow tests for estimating toxicity. *Water Res.* 22:883–893.

55. Giesy, J.P. and Graney, R.L. 1989. Recent developments in and intercomparisons of acute and chronic bioassays and bioindicators. In: M. Munawar, G. Dixon, C.I. Mayfield, T. Reynoldson, and M.H. Sadar, Eds., *Environmental Bioassay Techniques and Their Application*, Kluwer Acad. Publ., Dordrecht. pp. 21–60.
56. Dunnett, C.W. 1955. Multiple comparison procedure for comparing several treatments with a control. *J. Amer. Statist.* 50:1096–1121.
57. Dunnett, C.W. 1964. New table for multiple comparisons with a control. *Biometrics* 20:482–491.
58. Williams, D.A. 1971. A test of differences between treatment means when several dose levels are compared with zero dose control. *Biometrics* 27:103–117.
59. Williams, D.A. 1972. The comparison of several dose levels with a zero dose control. *Biometrics* 28:519–531.
60. Anderson, S.L. and Norberg-King, T.J. 1991. Precision of short-term chronic toxicity tests in the real world. *Environ. Toxicol. Chem.* 10:143–145.
61. Gelber, R.D., Lavin, P.T., Metha, C.R. and Schoenfeld, D.A. 1985. Statistical analysis. In: G.M. Rand and S.R. Petrocelli, Eds., *Fundamentals of Aquatic Toxicology,* Hemisphere Publ., Washington, D.C., pp. 110–123.
62. Hoar, W.S., Randall, D.J., and Brett, J.R., Eds. 1979. Bioenergetics and growth. *Fish Physiology*, Vol. VIII. Academic Press, 786 pp.
63. Wagner, G.F. and McKeown, B.A. 1985. Cyclical growth in juvenile rainbow trout, *Salmo gairdneri. Can. J. Zool.* 63:2473–2474.
64. Neilson, A.H., Allard, A.-S., Fischer, S., Malmberg, M., and Viktor, T. 1990. Incorporation of a subacute test with zebra fish into a hierarchical system for evaluating the effect of toxicants in the aquatic environment. *Ecotox. Environ. Safety,* 20:82–97.
65. DeGraeve, G.M., Cooney, J.D., McIntyre, D.O., Pollock, T.L., Reichenbach, N.G., Dean, J.H., and Marcus, M.D. 1991. Variability in the performance of the seven-day Fathead minnow (*Pimephales promelas*) larval survival and growth test: An intra- and interlaboratory study. *Environ. Tox. Chem.* 10:1189–1203.

CHAPTER 10

Standardizing an Aquatic Microcosm Test*

Frieda B. Taub

ABSTRACT

A Standardized Aquatic Microcosm (SAM) consisting of chemically defined medium and sediment, 10 species of algae, and 5 species of animals has undergone a round-robin test in 3 laboratories and the host laboratory. The experimental design consisted of treatment groups of replicates each of Control, 500, 1000, and 2000 µg/l Cu as $CuSO_4$.

Results in the different laboratories showed the same trends. In the Controls, nutrient depletion occurred as the algae increased; *Daphnia* populations increased and over-grazed the algae. In microcosms treated with 500 µg/l Cu, algal growth was inhibited only briefly, *Daphnia*

* School of Fisheries Publication 870. This manuscript was prepared while the author was a National Research Council Senior Resident Associate at U.S. EPA, Athens, GA Research Laboratory.

0-87371-845-3/93/$0.00 + $0.50

were temporarily eliminated, and during this period, algal concentrations were much greater than controls. *Daphnia* re-established populations from reinoculated animals, and eliminated the algal bloom. At 1000 μg/l Cu, algal growth was suppressed for a longer period of time, but recovered in all experiments; *Daphnia* were eliminated in all tests, and eventually recovered in most. At highest Cu concentration, 2000 μg/l, the algal suppression occurred for a greater duration and recovery occurred in only some experiments; *Daphnia* recovered during the 63-d experiment in only a few replicates of one experiment.

Seven experiments (two from the host laboratory, two each from two participating laboratories and one from a third participating laboratory) were compared statistically. Trends (comparing each treatment with its control) were statistically similar in all experiments. The results were similar enough to come to the same conclusions; however, the experiments were different from each other if variables were compared at each sampling date because of differences in timing of blooms and recovery from toxicity between experiments.

The major problems of reproducibility were associated with *Daphnia* rearing conditions and over-inoculation of blue-greens. These problems were experienced only in the host laboratory when *Daphnia* rearing techniques were varied from those used in the original development of the protocol. A synthetic medium (T85MVK) was developed that was satisfactory if local, natural waters are not available to support *Daphnia* culture.

KEY WORDS

Microcosms, multispecies bioassays, ecotoxicology, *Daphnia*, algae.

INTRODUCTION

Having a standardized, community level ecotoxicological test could make ecological properties, such as algae-zooplankton-decomposer interactions and dominance shifts within trophic levels, more acceptable for regulatory decisions concerning the potential safely of chemical releases into the environment.[1] There has been a reluctance to use complex community ecotoxicological tests as long as such tests are feared to be non-reproducible or subject to variable interpretations. Allowing or forbidding the introduction of chemicals into the environment involves expensive decisions. Ecological data would be more acceptable if there were assurances that results obtained in one laboratory could be reproduced in another.

The objectives of the research were to provide clear and practical experimental procedures that could be performed by other laboratories and would result in: (1) reproducible control responses similar to the nutrient-algae-grazers-recycling interactions shown in temperate aquatic communities, (2) sufficient replication within an experiment to demonstrate statistical differences caused by treatments, and (3) sufficient reproducibility between experiments to provide similar conclusions. The experimental procedures include quality control criteria for rejection of the results of a flawed experiment.

The word "Standardized" in Standardized Aquatic Microcosm (SAM) was used to imply that each experiment would be initiated with the same chemical solution and sediment, the same species, and to the extent feasible, the same initial age structure, population numbers, and physiological states. Like standardized single species tests, the intent was to allow comparisons of chemicals, while keeping the biological components as similar as practical. Experience had shown that the control behavior could provide a standard pattern against which treatment effects in the same experiment could be compared. Control behavior could also be a means of comparing between experiments. When the same concentrations of the same chemical were tested in different laboratories, the interlaboratory effects could be compared.

OVERVIEW OF EXPERIMENTAL METHODS AND RESULTS PRIOR TO THE INTER-LABORATORY TESTING

Each microcosm consisted of 3 l of chemically defined medium that satisfied the growth requirements of the 10 species of algae inoculated; 5 species of animals grazed on the algae or resultant detritus. Initial conditions were selected to be similar to early spring in temperate lakes and ponds;[2,3] algal nutrients were initially present at high concentrations, algae were initially inoculated at low initial conditions, but grew rapidly and depleted the dissolved nutrients as the algal bloom developed. Animal populations were introduced at low concentrations, but rapidly reproduced and grazed down the algal blooms. After the initial nitrate was depleted, continued algal growth relied on ammonia and phosphate recycled during grazing and decomposition.

Considerable study was done prior to the inter-laboratory testing. Earlier publications that can be consulted for additional detail are: the history of its development and the justification for the species selected;[4] the basis of sampling design and statistics:[5,6,7] effects of induced mortalities[8] or chemical effects of malathion (increased algal blooms as a result of grazer reductions), streptomycin (reduced grazers as a result of primary production inhibition), triethylene glycol (altered pH as a probable result of degradation) and copper (complex interactions including inhibition of algal growth and *Daphnia* mortalities);[9–11] comparisons of microcosms results with natural systems,[12] mathematical modelling of effects;[13–16] interlaboratory tests;[11,17–19] process studies that aided in the interpretation of microcosm or model results,[4,8,20–23] and generalized reviews.[10,24–26] The food chain structure and the timing of the toxicant addition were confirmed as being major controlling factors in responses by experiments with microcosms that included fish[27] or added the test chemical late at different times.[21,28]

Formal experimental protocols are available as ASTM E1366-90[29] and an FDA final report.[30] The latter is similar to the document that was used in the inter-laboratory testing exercise, and was used for a 1988 SETAC short course; it contains a supplies list and an introduction to a data handling program that uses EXCEL (Registered Trademark of Microsoft) and can be run on MS-DOS and Macintosh microcomputers.

This chapter presents the results of an inter-laboratory test including this (the host laboratory) and three other laboratories. Additional experiences conveyed to me by others who have used the protocol are also included.

PROTOCOL

In brief, a Standardized Aquatic Microcosm experiment[29,30] consists of approximately 24 microcosms (each of 3 l of chemically defined medium and 200 g of sand and 0.5 g each of chitin and cellulose, Table 1). On Day 0, 10 algal species (Table 2) are each inoculated at initial concentrations of 10^3 cells/ml after removal of nutrient samples for analyses. On Day 4, animals are added (Table 2) after the algae have

Table 1
Microcosm Media (T82MV) and Sediment Composition

Compound	Medium T82MV composition: Molecular weight	Concentration: mM	Element	mg/l
$NaNO_3$	85.0	0.5	N	7.0
$MgSO_4 \cdot 7H_2O$	246.5	0.1	Mg	2.43
KH_2PO_4	136.0	0.04	P	1.23
NaOH	40.0	0.099	Na	2.27
$CaCl_2 \cdot 2H_2O$	147.0	1.0	Ca	40.0
NaCl	58.5	1.5	Na	34.5
$Al_2(SO_4)_3 \cdot 18H_2O$	666.5	0.0048	Al	0.26
$Na_2SiO_3 \cdot 9H_2O$	284.0	0.80	Na	36.8
			Si	22.4
	Trace metals μM			
$FeSO_4 \cdot 7H_2O$	278.0	1.12	Fe	0.0625
EDTA	292.0	1.42	EDTA	0.4145
H_3BO_3	61.8	0.75	B	0.008
$ZnSO_4 \cdot 7H_2O$	287.5	0.025	Zn	0.0015
$MnCl_2 \cdot 4H_2O$	197.9	0.25	Mn	0.0135
$Na_2MoO_4 \cdot 2H_2O$	242.0	0.025	Mo	0.0024
$CuSO_4 \cdot 5H_2O$	249.7	0.005	Cu	0.00032
$Co(NO_3)_2 \cdot 6H_2O$	291.0	0.0025	Co	0.00015
	Murphy's vitamins μM			
Calcium pantothenate	476.5	1.47		0.70
Cyanocobalamin (B_{12})	1355.4	0.000022		0.00003
Thiamin (B_1)	337.3	0.18		0.06
Riboflavin (B_2)	376.4	0.11		0.04
Nicotinamide	122.1	1.06		0.13
Folic acid	441.4	0.75		0.33
Biotin	244.3	0.12		0.03
Putrescine	161.1	0.19		0.03
Choline	181.7	2.75		0.50
Inositol	216.2	5.09		1.10
Pyridoxine monohydrochloride	205.7	2.43		0.50

Note: pH adjusted to 7.0 with sterile 10% HCl. Sediment composition: 200.0 g silica sand. 0.5 g chitin. 0.5 g cellulose powder.

From Taub, F. B. and Kindig, A. C., Final Report, Food and Drug Administration Contract No. 223-83-7000, Section II, 1986.

been enumerated and nutrient samples removed. The initial animal numbers are: Protozoa, 0.1 per ml; rotifers 0.03 per ml; 16 *Daphnia* (3 large with eggs, 3 without eggs, and 10 small); 12 amphipods (6 adults and 6 small); and 6 Ostracods. Reinoculation is performed weekly in which small numbers of microscopic organisms (algae, rotifers, and

Table 2
Microcosm Organisms

Organism	Function
	Primary Producers
Anabaena sp.	Blue-green, filamentous, potential nitrogen fixer, potential amphipod food. 10 μm³ per cell, filaments about 290 μm³, assumed only 20% of the cells are available to *Daphnia*
Ankistrodesmus sp.	Green, unicellular, 43 μm³ per cell, assumed 80% available to *Daphnia*
Chlamydomonas reinhardi (90)	Green, unicellular, 22 μm³ per cell, assumed 100% available to *Daphnia*
Chlorella vulgaris	Green, unicellular, 22 μm³ per cell, assumed 100% available to *Daphnia*
Gomphonema parvula	(Diatom 213), unicellular, 48 μm³ per cell, assumed 70% available to *Daphnia*
Nitzschia kutzingiana	(Diatom 216), unicellular, 51 μm³ per cell, assumed 70% available to *Daphnia*
Lyngbya sp.	Blue-green, filamentous, not a nitrogen fixer, potential amphipod food, 10 μm³ per cell, about 695 μm³ per filament, assumed 1% available to *Daphnia*
Scenedesmus obliquus (72)	Green, colonies of four, 133 μm³ per cell, assumed 66% available to *Daphnia*
Selanastrum capricornutum	Green, unicellular, 22 μm³ per cell, assumed 100% available to *Daphnia* (also, commonly used for bioassays)
Stigeoclonium sp.	Green, filamentous, available to amphipods, 20 μm³ per cell, filaments about 206 μm³, assumed 1% available to *Daphnia*
Ulothrix sp.	Green, filamentous, available to amphipods, 185 μm³ per cell, filaments exceed 300 mm³, assumed 2% available to *Daphnia*
	Secondary Producers
Cyprinotus incongruens or *Cypridopsis*	Ostracod, feeds on detritus, especially small particulates
Daphnia magna	Cladoceran, feeds on algal cells and other particulates
Hyallela azteca	Amphipod, feeds on filamentous algae and detritus, suspected of feeding on small *Daphnia*
Philodina sp.	Rotifer, feeds on bacteria and small algal cells, other small particulates
Small rotifer	Rotifer (smaller than *Philodina*), presumably similar feeding habits
Hypotrichs	Protozoan, a large ciliate, presumably feeds on very small particles and on dissolved organics
Other protozoa	Several types introduced with the *Daphnia*, amphipods, and ostracods, the numbers and types are monitored
Bacteria and fungi	Undoubtedly introduced with the other organisms, not routinely monitored but could be enumerated

From Taub, F.B., et al., *Aquatic Toxicology and Hazard Assessment: Sixth Symposium*, ASTM STP 802, Bishop, W.E., Cardwell, R.D., and Heidolph, B.B., Eds., American Society for Testing and Materials, Philadelphia, 1983, 5. With permission.

protozoa) are added. These organisms are added after the organism counts; the numbers of microscopic organisms are so small that they would be below detection if reproduction did not occur. If the *Daphnia,* ostracods or amphipods are fewer than 3 organisms per microcosm, enough are added to achieve 3. The purpose of the reinoculation is to reduce the potential for random extinctions and to allow community recovery from temporarily toxic conditions. It is similar to immigration in natural systems.

On Day 7, the 24 most similar microcosms are selected on the basis of changes in dissolved oxygen from a.m. to p.m., pH, number of *Daphnia,* and numbers of *Selenastrum* and *Chlamydomonas.* Cracked microcosms are eliminated because small cracks tend to grow. The eliminated (culled) microcosms can be used for other purposes, such as culturing amphipods and ostracods. The 24 most similar microcosms are randomly assigned to treatment groups; in this example, the treatment groups consist of 6 replicates each of control, low, medium, and high concentrations of a test chemical (Cu as $CuSO_4$). Other experimental designs may be appropriate depending on the purpose and statistical analysis to be used,[29,30] but statistical precautions should be considered.[5]

The variables measured included twice weekly enumeration of organisms, pH, oxygen concentrations (just before lights-on, late afternoon, next day just before lights-on), optical density, and *in vivo* fluorescence; twice weekly for the first month and then weekly, nutrient analyses of phosphate, nitrate, nitrite, and ammonia. From the measured variables, others were calculated, such as dissolved oxygen gain (a measure of net daytime photosynthesis) which was calculated from the difference between the first two oxygen measurements (Table 3). Sampling procedures do not require aseptic technique; apart from maintaining the stock algal, protozoan, and rotifer cultures, aseptic technique is not used. The amphipods, ostracods, and *Daphnia* cultures include undefined microbiota.

The enumeration of organisms and inverse sampling technique were used, in which larger volumes of water were searched for larger or rarer species. For example, in the process of counting algal cells on a Palmer slide, cells of all genera were counted in the first microscopic field; thereafter, as soon as 50 cells of a genus were counted, the number of cells and number of fields counted were recorded and that genus was no longer counted. Thus, common genera were counted in only a few microscope fields and only rarer species were counted in as many as 65 fields (the capacity of the Palmer slide). For large organisms such as

Table 3
Variables Measured in a Standardized Aquatic Microcosm

	Immediate data	
1	SELAN	*Selenastrum* × 10^4 cells per ml
2	CHLAMY	*Chlamydomonas* × 10^4 cells per ml
3	SCENED	*Scenedesmus* × 10^4 cells per ml
4	ANKIS	*Ankistrodesmus* × 10^4 cells per ml
5	CHLOR	*Chlorella* × 10^4 cells per ml
7	NITZ216	*Nitzschia kutzigiana* × 10^4 cells per ml
8	ANABENA	*Anabaena* × 10^4 cells per ml
9	LYNGBYA	*Lyngbya* × 10^4 cells per ml
10	STIG	*Stigeoclonium* × 10^4 cells per ml
11	ULOTH	*Ulothrix* × 10^4 cells per ml
12	SDAPHNIA	Small *Daphnia* per 100 ml
13	MDAPHNIA	Medium *Daphnia* per 100 ml
14	LDAPHNIA	Large *Daphnia* per 100 ml
15	DAPHNIA	*Daphnia* per 100 ml all sizes
16	PSDAPHN	Ratio of small to total *Daphnia* (optional)
17	MDSIZE	Mean *Daphnia* length (optional)
18	DBIOMAS	*Daphnia* biomass per 100 ml (optional)
19	SAMPH	Small amphipods per 100 ml
20	LAMPH	Large amphipods per 100 ml
21	AMPH	Amphipods per 100 ml
22	OSTRACOD	Ostracods per 100 ml
24	EPHIPPIA	Ephippia presence
25	PHILODIN	*Philodina* per ml
26	ROTIFER	Small rotifers per ml
27	HYPO	Hypotrichs per ml
28	MISCEL	Miscellaneous protozoans per ml
29	VORT	*Vorticella* per ml (if observed)
30	NEMER	Nemertines per ml (optional)
33	ABSORB	Absorbance
34	DO1	Dissolved oxygen before lights on (ppm)
35	DO2	Dissolved oxygen before lights off (ppm)
36	DO3	Dissolved oxygen before lights on (ppm)
37	DDOAM	Dissolved oxygen gain (ppm)
38	DOOPM	Dissolved oxygen loss (ppm)
39	DELDO	Dissolved oxygen change 24 h (ppm)
40	RATIOPR	Net photosynthesis-respiration ratio
41	FLUOR	Fluorometry
42	PHL	pH Immediately before lights on (0930 hours)
43	AVALGAE	Available algae
44	ABIOVOL	Algal biovolume × 10^4 cells cubic microns per ml
47	DFECUND	*Daphnia* fecundity ratio (optional)
49	DVRSFD	Shannon weaver weighted algal diversity
50	DVRSALG	Shannon weaver algal species diversity
	Nutrients	
51	PO_4	Phosphate
52	SIO_4	Silicate
53	NO_3	Nitrate
54	NO_2	Nitrite

Table 3 (cont.)

	Nutrients	
55	NH_3	Ammonia
56	RATIO1	Nitrate-phosphate atomic ratio (optional)
57	RATIO2	Nitrate + nitrite-phosphate atomic ratio (optional)
58	RATIO3	Total N-P atomic ratio (optional)

From Taub, F. B., in *Concepts in Marine Pollution Measurements,* White, H. H., Ed., University of Maryland Sea Grant College, 1984, 159. With permission.

Daphnia, amphipods, and ostracods, if only a few organisms were present, they were counted visually in the entire microcosm. If there were too many to count in this manner, 100 ml sub-samples were removed; the number of sub-samples was adjusted to assure that at least 20 specimens of each genus were sampled. The computer calculated the algal cells per ml and the animals per 100 ml from the numbers and volumes counted. These inverse sampling procedures allowed better estimates of rare organisms without excessive effort spent on abundant organisms. Species enumeration data were log transformed to reduce variance.

Quality control criteria include provisions for evaluating the quality of an experiment on Day 28, and throughout the entire experiment (to Day 63). If by Day 28, the nitrate has not been reduced from the initial concentration of 0.5 m*M* to less than 0.05 m*M*, the algal volume has not achieved a mean peak volume of more than 2000×10^4 cubic microns/ml, the oxygen gain has not achieved a mean peak value of more than 4 mg/l sometime during the period, the *Daphnia* has not achieved an abundance of more than 100 individuals per 100 ml, it was recommended that the experiment should be terminated. The formal protocols also set Coefficient of Variation (C.V.) and pH limitations. If these criteria were not met, it is likely that the medium was made incorrectly or the organisms were unhealthy. During the remainder of the experiment, control mean oxygen gain must be positive, the *Daphnia* should be more abundant than 15 individuals per 100 ml, and the lower bound of the interval of non-significance must be greater than 0 for more than 50% of the sampling period after the introduction of the test material in the treated microcosms. By setting a limit to the lower bound, the experiments would be unacceptable if extinction of *Daphnia* in a treatment group could not be shown to be significantly different from the controls. The only rejection criteria that affected experiments reported here was the requirement for *Daphnia* populations toward the end of some experiments.

PARTICIPATING LABORATORIES — INTERLABORATORY TESTS AND OTHERS

In addition to our laboratory (University of Washington), the participating laboratories were (1) EPA-Duluth (with a cooperative agreement with the University of Minnesota at Duluth, MN); (2) U.S. Army, Aberdeen Proving Ground, MD; and (3) Marine Bioassay Laboratory, Carlsbad, CA (all U.S.). The intent was to have each laboratory conduct 2 microcosm experiments, each to consist of 0, 500, 1000, and 2000 μg/l Cu as $CuSO_4$. Marine Bioassay Laboratory was only able to conduct one experiment, and the first of the U.S. Army, Aberdeen Proving Ground experiments did not use 1000 and 2000 μg/l. Within our laboratory, we conducted 5 experiments, 2 of which were used in the interlaboratory comparison. Thus, we had 7 experiments for the controls and lowest concentration and 6 experiments for the two higher concentrations (Table 4).[5,10,11,17–19]

Prior to the initiation of the interlaboratory test, each of the participating organizations sent scientific staff to the host laboratory for a one week training session. We discussed the procedures in the manual[30] and demonstrated the more difficult techniques. The visiting scientists practiced identifying, counting, and recording the organisms, measuring oxygen and pH (prior to lights-on), etc. Each organization left with the manual, a set of organisms, a small sample of our medium, and small samples of our medium master solutions.

To test for the quality of the distilled water and master solutions used in the microcosm experiments, we asked each participating laboratory to demonstrate its ability to grow algae and maintain the animals in (1) media made in our laboratory with our master solutions and distilled water, (2) media made from our master solutions diluted with their distilled water, and (3) media made entirely in their laboratory.

The data were recorded by each participating laboratory and a copy sent to us for calculations and within-experiment statistical analysis on our computerized-data handling-package and mainframe CDC computer. The calculated tables, graphics, and statistics were sent to participating laboratories for confirmation that they represented the data submitted.

During the period that the participating laboratories were conducting their experiments, we conducted 5 experiments. We modified our methods and rewrote the manual[30] to eliminate problems that were identified.

Table 4
Interlaboratory Testing Experiments

Laboratory	First experiment	Second experiment
Accepted experiments (included in DWV statistical comparison)		
University of Washington	UW 1R	UW 2R
Initiated	May 2, 1986	Jan. 24, 1986
Exper. no.[a]	ME 89	ME 82
Cu deviation[b]	+10%	+5%
EPA-Duluth	DU 1	DU 2
Initiated	Dec. 2, 1983	April 6, 1984
Exper. no.	ME 65	ME 67
Cu deviation	–3%	+6%
Aberdeen-Army	AB 1	AB 2
Initiated	May 3, 1985	Jan. 24, 1986
Exper. no.	ME 74[c]	ME 87[d]
Cu deviation	3	–10%
Marine Bioassay Laboratory	MB 1	
Initiated	Nov. 15, 1985	
Exper. no.	ME 79	
Cu deviation	–1%	
Rejected experiments (documented)		
University of Washington	UW 1 (original)	UW 2 (original)
Initiated	Dec. 2, 1983	Oct. 12, 1984
Exper. no.	64	69
Cu deviation	–13	+18%
University of Washington	—	
Initiated	Feb. 1, 1985	
Exper. no.	72	
Cu deviation	+25%	

[a] Experiment numbers indicate the model ecosystem number under which the original data are stored in the host laboratory.
[b] Cu deviation indicates the difference between the nominal concentration of Cu and the measured total Cu on Day 7.
[c] For this experiment only, the treatments consisted of control, 127, 255, and 509 μg Cu per liters; all other experiments consisted of control, 500, 1000, and 2000 μg Cu per liter.
[d] Nutrient analyses are incomplete for this experiment at this time.

From Taub, F. B., Kindig, A. C., Conquest, L. L., and Meador, J. P., in *Eleventh ASTM Symp. Aquatic Toxicology and Hazard Assessment,* ASTM STP 1007, Suter, G. W., II and Lewis, M. A., Eds., American Society for Testing and Materials, Philadelphia, 1988, 368. With permission.

After the completion of the formal inter-laboratory test, R. R. Lintott of the University of Saskatchewan and Dr. S. Swanson of Beak Associates, Saskatoon, Saskatchewan used the method with radiolabeled diclofop-methyl.[31] Other studies have been performed at Aberdeen Proving Ground.[23] Suggestions made by these users were incorporated in the ASTM manual.[29]

MEDIA MAKING AND ORGANISM CULTURES

In general, all of the laboratories experienced as good results with medium prepared in their own laboratories as medium taken from our laboratory or made with their distilled water or master solutions. Although there were some statistical differences between algae growth rates in the various media, the differences were not of great magnitude, and medium made entirely from the participating laboratories supplies did as well or better than medium or stock solutions from our laboratory. Animal survival was also good. None of the laboratories experienced "toxic distilled water" problems during these studies.

Standard rearing media for *Daphnia* and amphipods were more difficult to develop. Prior to the inter-laboratory test, we had reared our *Daphnia* in Lake Washington water, supplemented with the algae *Scenedesmus* and *Ankistrodesmus*. During the training workshop, our attention was drawn to our use of a "site-specific" water. Marine Bioassays Laboratory did not have a satisfactory local water supply for culturing *Daphnia*. Their use of de-chlorinized tap-water and algae resulted in *Daphnia* with low fecundities. We tested long term *Daphnia* culture in the microcosm medium (T82MV), but it was not satisfactory. After one or two generations, reproductive rate and general health of the animals were reduced; the second antennae were often abnormal, and protozoans were often attached to the carapace. Use of these *Daphnia* in microcosm experiments (ME64, 69, and 72) gave poor results. Also, in rearing *Daphnia* in the microcosm medium, active photosynthesis often produced high pH conditions that were detrimental to *Daphnia*.

We therefore developed a medium T85MVK to overcome the inadequacies of T82MV. The requirement of *Daphnia* for Se had been shown,[33,34] and it was possible that other trace metals might also be required. The new medium, T85MVK differed from the microcosm medium by having 1/10 the nitrate and phosphate concentrations, thus reducing photosynthesis by the algae, and having the salts recommended by Keating that supplied elements not already in the T82MV medium (Br, Sr, Rb, Li, I, Se, and V). *Daphnia* reared in T85MVK were healthier than those reared in T85MV, and gave adequate microcosm results (ME89) but these *Daphnia* were less healthy than those reared in Lake Washington water or in Lake Superior water in Duluth, MN. We recommend that local lake water be used, if convenient and available, and that T85MVK be used in the absence of a satisfactory natural water.

The amphipod rearing techniques used in our laboratory seemed adequate during the inter-laboratory tests, but were not satisfactory in Canada. We had reared amphipods in old control microcosms and fed them the filamentous algae *Ulothrix* sp. and *Stigeoclonium* from the semi-continuous algae cultures. Swanson found that amphipods collected in the natural environment required gradual exchange of their water with microcosm medium T82MV over a period of several weeks. Bottom-feeding fish food was found to be satisfactory as an alternative to algae. Light levels of 5000 to 7500 lux with a 18:6 light:dark cycle were satisfactory.[41]

The synthesis of the microcosm components differed between laboratories; we are not aware that this caused any differences in results. In our laboratory, having an autoclave that could handle 8 microcosm bottles at a time, we preferred to add the weighed quantities of washed sand, chitin, cellulose, and 3 l of "basal medium" (without phosphate, trace metals, and vitamins solution), cover with aluminum foil, and autoclave. The Aberdeen laboratory preferred to autoclave the bottles empty, and to autoclave all components, and combine them at the beginning of an experiment. In both methods, the phosphate, trace metals, vitamin solutions, and HCl (for pH adjustment to 7.0) were added aseptically at the beginning of the experiment.

Other modifications occurred when adequate quantities of a species were unavailable at the beginning of the experiment. Aberdeen omitted amphipods in both of their inter-laboratory experiments, and omitted *Ulothrix* in their second experiment. They did not measure *in vivo* fluorescence. Differences in nitrate-nitrite analyses occurred. Duluth used a combined nitrate-nitrite measurement in their first experiment. Combining the measurements has effect on treatment effects because nitrite concentration is very low except for a brief period when nitrate approaches depletion.

MICROCOSM RESULTS — BETWEEN LABORATORIES

Controls

The sequences of nutrient depletion, algal blooms, *Daphnia* blooms concurrent with reductions in algal volume occurred in all experiments

(Figure 1a, e, i).[11] Although there were differences between experiments, general patterns were similar.

The nominal initial nitrate concentration was 0.5 m*M*; measured concentrations on Day 0 ranged somewhat (from 0.424 to 0.573). The range of differences in initial nitrate concentration did not appear to have important differences on microcosm responses. In two experiments, the nitrate concentration increased from Day 0 to Day 4 (in UW#1 nitrate changed from 0.424 to 0.612; in DU#1 it changed from 0.518 to 0.556 m*M*); this increase may have been an analytical artifact associated with dilution required for these concentrations, or may have been caused by the addition of nitrate with the algal inoculum. Nitrification was not observed in these microcosms, and any increases were attributed to analytical errors (for example, the very high value on Day 31 in MB#1, 2000 µg/l Cu), or occasional slight increased in nitrate concentration.

The nitrate depletion occurred in all experiments, but occurred earlier in some experiments than others (Figure 1a). There were similar differences in the timing of the algal bloom; algal blooms occurred earlier in some experiments than in others (Figure 1e). This coupling of nitrate

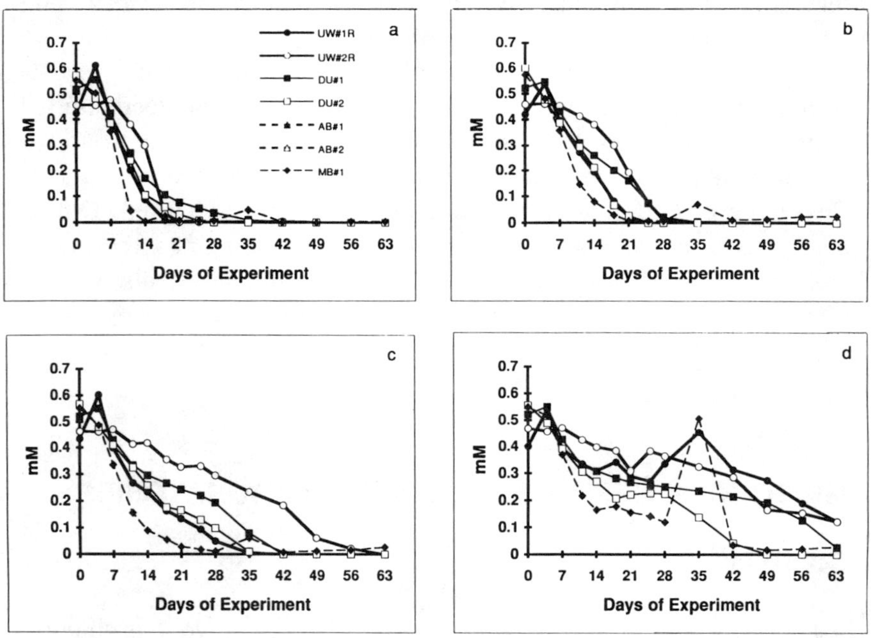

FIGURE 1. Interlaboratory reproducibility. (a-d) Nitrate in m*M* (mg-a+/l). Each line is the mean of 6 replicates. (From Taub, F.B., et al., in *Aquatic Toxicology and Hazard Assessment: Eleventh Symposium*, ASTM STP 1007, Suter, G.W., II and Lewis, M.A., Eds., American Society for Testing and Materials, Philadelphia, 1988, 368. With permission.)

depletion with the development of the algal bloom is to be expected because algal uptake is the mechanism by which nitrate is removed from the medium.

Daphnia population increases occurred in all experiments, with slight differences in timing and magnitude (Figure 1i). All of the *Daphnia* populations peaked between Days 20 and 30. Differences in the magnitude of the *Daphnia* population peaks were attributable, at least in part, to the water supplies available for rearing stock cultures. The Duluth *Daphnia* reached much greater peak densities than did the other laboratories; they had been reared in Lake Superior water. Marine Bioassay Laboratory *Daphnia* had lower reproductive rates and had low peak densities. This laboratory did not have a local water supply suitable for *Daphnia* rearing, and used aged (biologically conditioned) microcosm medium T82MV. Early in the experiment, the rate of *Daphnia* increase depends on the initial nutritional condition of the *Daphnia* inoculated. We had not standardized for the number of eggs carried by the inoculated adults, which provide the first offspring. Also, the initial nutritional state of the sub-adults can be expected to affect their early brood sizes.

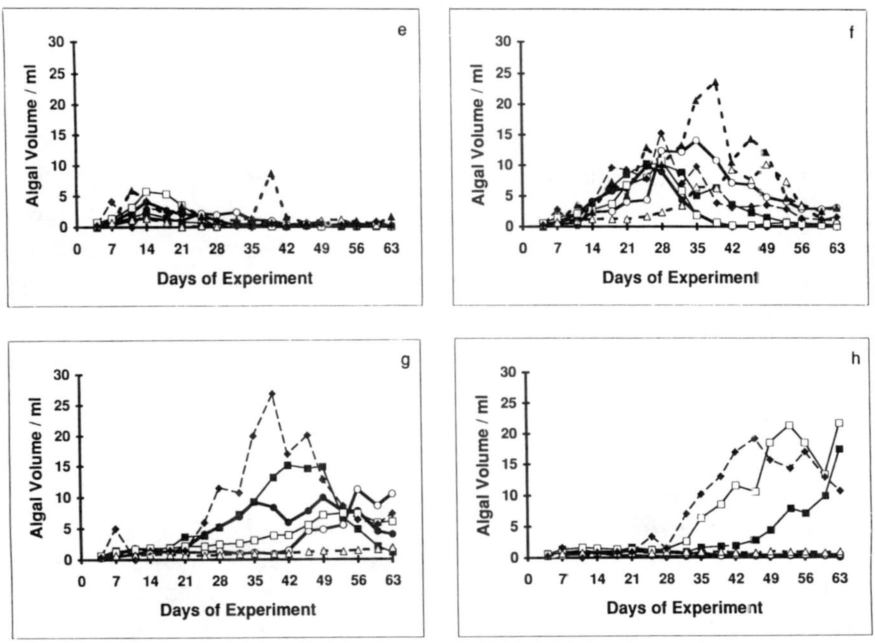

FIGURE 1 (e-h). Algal volume (10^7 cubic microns per ml). Each line is the mean of 6 replicates.

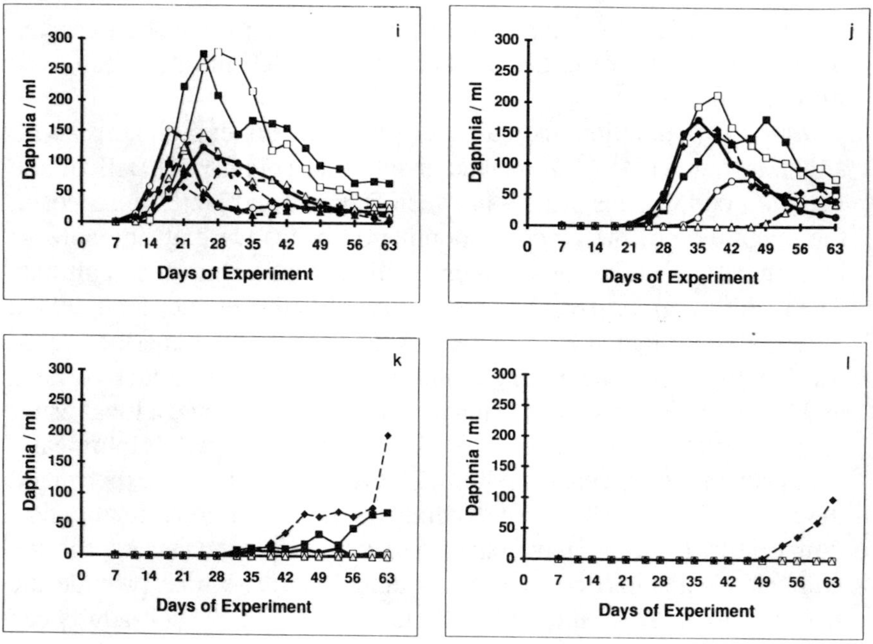

FIGURE 1 (i-l). *Daphnia* (individuals per ml). Each line is the mean of 6 replicates. Upper left (a, e, i) are control data; upper right (b, f, j) are 500 μg/l data; lower left (c, g, k) are 1000 μg/l data; lower right (d, h, l) are 2000 μg/l data.

Much of the variability in timing between experiments can be explained by trophic interactions (Figure 2). In experiments with rapid depletion of nitrate, e.g., MB#1, the development of the algal bloom was early and of great magnitude, associated with slow development of the *Daphnia* population (Figure 2a). In experiments with slow depletion of nitrate, e.g., UW#2R, the rate and magnitude of the algal bloom was limited by the rapid increase in *Daphnia* populations. Statistical analyses of single variables fail to account for these expected trophic linkages between variables.

Copper Effects

Copper concentration measured on Day 7 differed slightly from the nominal concentrations of 500,1000, and 2000 μg/l Cu. The differences ranged from 10% greater to 10% less than nominal for the experiments used in the inter-laboratory comparisons (Table 4). In one of the inter-

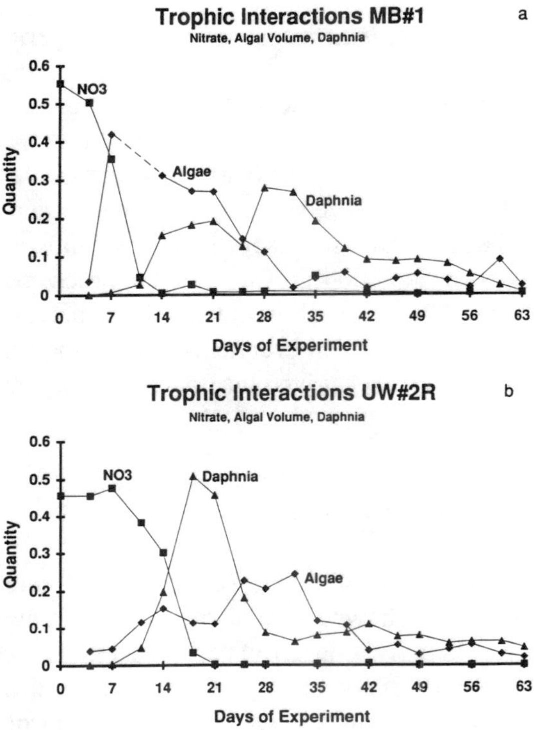

FIGURE 2. Trophic interactions among nitrate, algal volume, and *Daphnia.* (a) MB#1 a microcosm experiment with rapid nitrate depletion. (b) UW#2R a microcosm experiment with slow nitrate depletion. Units: Nitrate, μM; algae volume, $\times 10^6$ cubic microns/ml; *Daphnia* individuals/ml (divided by 3). The units are slightly modified from Figure 1 to ease comparison of variables in a single graph.

laboratory experiments, the directions had been interpreted as 500, 1000, 2000 µg/l $CuSO_4 * 5H_2O$; as a result that experiment had lower Cu concentrations; only the controls and its highest Cu concentration (which was approximately 500 µg/l Cu) could be compared. In some of their earlier experiments in our laboratory, the copper concentrations were 13% less to 25% greater than nominal (Table 4).

In every experiment, 500 µg/l Cu delayed the nitrate depletion, temporarily eliminated the *Daphnia*, and caused an excessive algal bloom during the absence of *Daphnia*, as compared to the control. *Daphnia* populations developed (from reinoculated organisms) in all 7 experiments.

At the next higher concentration of Cu (1000 µg/l), the algal populations were temporarily suppressed, but the algae eventually grew in all experiments, and the *Daphnia* recovered late in 2 experiments, but not in 4 others.

In the highest concentration of Cu (1000 μg/l), algae was suppressed in all of the microcosms for the first half of the experiments, and eventually recovered in 3 of the 6 experiments carried out at this concentration. *Daphnia* survived and reproduced only at the very end of the one of the six experiments. Meador[35] has since shown that recovery will eventually occur in 2000 μg/l Cu treated microcosms, if one waits as long as a year. The recovery is mediated by increases in pH and dissolved organic matter, which eventually chelate the Cu in non-toxic forms. After this occurs, sensitive algae and other organisms, such as *Daphnia* can become established.

Comparison of Conclusions

Although the comparison of different experiments (Figure 1) suggests that the 7 experiments showed similar trends, it is more important to compare the treatments within an experiment (References 11 and 5 provide statistical analyses in more detail). Because there are differences in the timing and magnitude among the control of different experiments, it is important that experimental treatments be compared with controls of the same experiment.

Would each of the experiments have resulted in the same conclusions being reached? Comparing the significant differences between controls and treatments within each experiment, each of the experiments demonstrated consistent results: copper delayed nitrate depletion, algal blooms and the development of *Daphnia* populations, and the greater the copper concentration, the greater the delay.[11]

To provide a statistical comparison of the copper effects among experiments, a statistic "Day-Weighted-by-Variable" was developed. The DWV statistic provides a measure of the lateral shift or phase lag. It is mathematically defined as:

$$\mathrm{DWV} = \frac{\Sigma(\mathrm{day} \times \mathrm{value})}{\Sigma(\mathrm{value})}$$

where value equals the organism abundance or nutrient concentration and where the summation is performed over all sampling days of a variable. The DWV is equivalent to computing a "center of gravity" in terms of the days for the time trace of each experimental variable. The DWV reflects differences between an early bloom (low DWV) and a period of inhibition followed by a later bloom (high DWV). The DWV values can be visually estimated from the data shown in Figure 1 by dividing the area under the curve into two equal areas and estimating the day on which the dividing line would fall — the later in the experiments this "center of gravity" occurs, the higher the DWV.

The calculated DWV values are shown in Figures 3a-c. The DWV values were compared by ranking Friedman's nonparametric test statistic.[11] The statistical analyses demonstrated that all experiments agreed that copper delayed the nitrate depletion, algal blooms, and *Daphnia* population development; the higher the copper concentration, the longer the delay in each experiment.

The actual amount of the delay with increased concentration varied enough between experiments to result in a significant treatment-by-treatment interaction term when subjected to a parametric two-way ANOVA. Thus, the trends were the same in all experiments, but the timing of responses differed. Differences in timing were especially obvious when recovery was involved. Since the recovery depended on biological activity (high pH and DOC),[35] conditions for detoxifying Cu occurred at different times. In the 2000 μg/l Cu treated microcosms, those that became contaminated by *Oocystis* (a Cu resistant alga) recovered faster.

WITHIN OUR LABORATORY

Although unplanned, our laboratory experienced a staff reduction prior to the inter-laboratory testing and many of the tasks in our laboratory were performed by less experienced personnel than had performed the experiments that led to the initial protocol development. We were also varying the methods of rearing *Daphnia*. Prior to the inter-laboratory experiments, our *Daphnia* had been reared in autoclaved Lake Washington water; our first three inter-laboratory experiments (ME 64, 69, and 72) used *Daphnia* reared in microcosm medium T82MV. Non-standard, aberrant results occurred in the controls of these experiments and the results were not included in the inter-laboratory statistical analyses, but are discussed here and elsewhere.[11,17] *Daphnia*

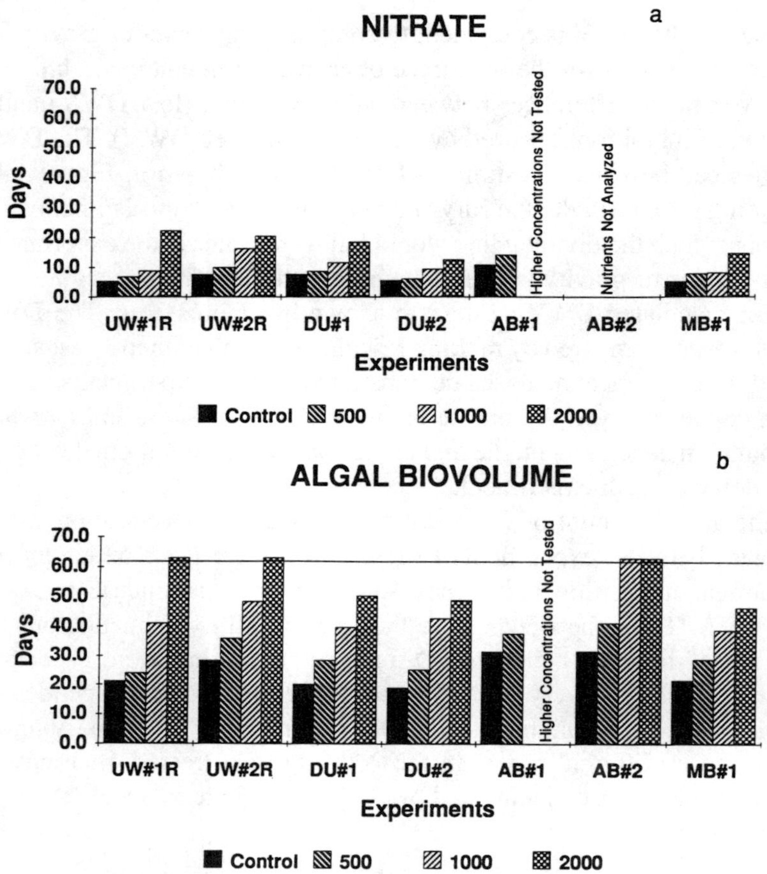

FIGURE 3. Comparison of results in seven SAM experiments. The Day-Weighted-by-variable statistic is a measure of the central distribution of the variable; a higher number indicates a delay. See text for additional discussion. (a) Nitrate; (b) Algal volume (if the bar extends above the day 63 line, populations had not become established by day 63.) (From Taub, F.B., et al., in *Aquatic Toxicology and Hazard Assessment: Eleventh Symposium,* ASTM STP 1007, Suter, G.W., II and Lewis, M.A., Eds., American Society for Testing and Materials, Philadelphia, 1988, 368. With permission.)

population development was poor and blue-greens became the algal dominants, whereas the usual dominants were green algae, usually *Ankistrodesmus* as the major dominant, with *Scenedesmus* as a much less dense, secondary dominant. When these aberrant results occurred in the first experiment (ME 64), we concluded that blue-greens probably had been inoculated in excessive amounts because clumps were not removed prior to dispensing the inoculum. We hypothesized that excessive inoculation of blue-greens and their dominance had adversely

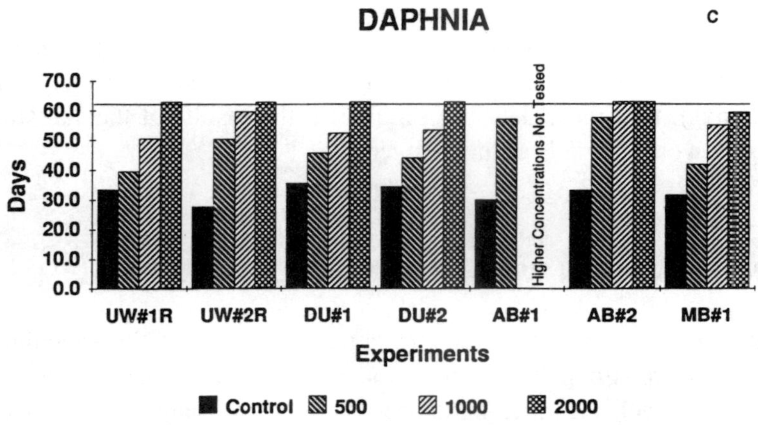

FIGURE 3C. *Daphnia* (if the bar extends above the day 63 line, populations had not become established by day 63.)

affected the *Daphnia.* However, similar results occurred in the next two experiments, in spite of care to inoculate appropriate amounts of blue-greens. Also, the dominance of blue-greens often seemed to follow, rather than precede, the poor *Daphnia* survival and reproduction. An experiment was performed to determine if Se deficiency stimulated the dominance of blue-greens, but the results were not supportive of that hypothesis. The only factors that these three experiments seemed to have in common was that the *Daphnia* stocks had been cultured in microcosm medium T82MV.

The experiments that failed to meet the quality control criteria failed because the *Daphnia* populations did not persist at sufficient densities in the controls to have the ability to demonstrate negative effects in the treatments. That is, the Interval of Significant Difference included 0. The rejected experiments demonstrated the same responses to copper. Data from one of the rejected experiments were reported in Taub et al.[17]

Use of *Daphnia* reared in autoclaved Lake Washington water (ME 82) provided adequate *Daphnia* population development and satisfactory microcosm results (shown in this chapter). Review of the literature indicated that Se and other trace metals might be necessary.[33,34] We then developed T85MVK with the addition of Keating's salts as a rearing medium for *Daphnia*; *Daphnia* raised in this medium were also satisfactory for microcosm studies (ME 89, results shown in this chapter). These two experiments, ME 82 and 89, were used in the inter-laboratory statistical comparison.

We did not include the three rejected UW experiments in the inter-laboratory statistical comparison because we did not want to use more experiments from our laboratory than from each of the participating laboratories (two each from Duluth and Aberdeen, and one from Marine Bioassays). Had these data been included, the results of the statistical analysis would have been unchanged.

CONTAMINATION WITH NON-INOCULATED ANIMALS

Numerous protozoans were present in addition to the inoculated Hypotrich protozoan. In a few cases, *Daphnia* had large numbers of *Vorticella* attached to their carapaces. These protozoans did not appear to modify results.

The algae *Oocystis* became established in the Duluth microcosms that had been treated with high concentrations of copper. These were the first algae to recover from toxicity. These *Oocystis* were isolated and sent to our laboratory for testing; they seemed to be inhibited at modest levels of Cu, but at higher levels, e.g., 2000 to 5000 μg/l, they grew better than controls in T82MV medium (unpublished results). *Oocystis* "invaded" our high-treatment microcosms in our last experiment (ME 89); they were the first algal species to show growth. The dissolved organic carbon and elevated pH caused by *Oocystis* was important in detoxifying the microcosms for other algae, and eventually for the survival and reproduction of *Daphnia*.[35] *Oocystis* never established dense populations in control microcosms or those treated with 500 μg/l Cu.

DISCUSSION

It had once been hypothesized that simple systems, lacking the complexity of natural systems, would tend to develop in unpredictable and non-reproducible ways. In fact, their behavior is highly predictable, although the exact timing varies somewhat even within an experiment. The degree of replication for these and other experiments have been published.[4,8,9–11,17–19,21,22,25,26]

Given that every additional step of standardization increases the cost of experiments, I suggest only limited improvements in standardization. I would recommend that the *Daphnia* rearing methods be monitored, and that the number of eggs per female be controlled so that each experiment would tend to have similar numbers of newborn *Daphnia* 2 to 4 d after the *Daphnia* are inoculated. We had shown earlier that *Daphnia* is a critical species, and its loss or marked reduction has impacts on many variables.[8,23] Also, I would recommend that the initial algal inocula be well mixed and clumps be eliminated so that filamentous algae not be over-inoculated. The initial algal concentrations will influence the timing of events,[22] and small initial differences may be a cause of increased variance between replicates and between experiments.

The greatest shortcoming in multi-species test is the statistics available. All that we have tried thus far fail to consider the known interactions between trophic levels. We know that more rapid development of grazer (*Daphnia*) population is associated with more rapid reduction in algal abundance; but the statistics compare only single variables on measurement days (e.g., nutrients, algae, *Daphnia* concentration). The same problem exists with comparisons between experiments; if algal growth should occur faster in one experiment than another, it will be associated with faster depletion of algal nutrients, and faster development of the *Daphnia* populations (with concurrent changes in oxygen and pH). Using most statistical comparisons of concentrations on a specific day, all variables will be statistically different between experiments. We do not have an *a priori* statistic that will state that the results were similar, but that events happened a sampling day earlier in the second experiment. In spite of our having built in the trophic level relationships, and having a mechanistic simulation model,[13–16] we have not succeeded in incorporating these into statistical comparisons between systems.

There are many pros/cons of single species vs. microcosm experiments. Almost always, knowing the sensitivities of the component species is useful in understanding if an event in the microcosms is caused by direct or indirect causes. Some species interactions could be predicted from having single species sensitivities; e.g., an experienced ecologist could predict that algal blooms might occur if grazers are poisoned. However, our ability to understand how species competition is affected by toxicants is limited; e.g., in the sensitivities of various algae to streptomycin, it was easy to predict that blue-greens would be eliminated and some greens would have competitive advantage. However, among the greens, it was not easy to determine the competitive outcome entirely on the basis of a single species test. For both streptomycin and copper, single species showed different sensitivities depending on such factors as cell densities, pH, exposure time, etc. It was not a simple matter to predict, based on single species toxicity tests, which species would be the competitive winner in the microcosm.[9,15,20]

A major advantage of the interactive communities is that microcosms allow the presence of the same organisms to alter the toxicant, and thereby affect the response of other organisms. TEG was an example; after a month's time, its addition to microcosms caused acidic conditions that altered the ability of other organisms to survive.[10] In the case of Cu, the presence of Cu resistant algae allowed detoxication via DOC and pH and the survival of sensitive organisms.[11,17,18,28,35] In this sense, the microcosms may provide more realistic results than single species tests.

The ideal number of organisms that should be present in a multispecies test depends on the purpose. In natural systems, there seems to be an almost infinite number of non-dominant species that can become dominant, if conditions reduce the competitive advantage of the present dominant. These shifts in species dominance occur regularly in seasonal planktonic succection. Most ecological communities respond to moderate toxic or other stresses by shifts in the species present, rather than by loss of all species, or proportional reductions in size or health of the species present.[36] Thus, some reasonable degree of trophic level redundancy seemed a requirement of a system that could mimic the properties of natural communities.

We had originally based the need for 10 algal species on the criteria of having several genera of algal phyla — blue-greens, greens, and diatoms. We were somewhat limited by the inability of several diatoms, which grew very well in the microcosm medium T82MV as unialgal cultures, to compete and establish measurable populations in the presence of the mixed algal biota. At the time we initiated our methods development, species diversity was thought to be a very important property of ecological systems. The data handling package automatically calculates species diversity of the total algae and that available to *Daphnia* (Shannon Weaver diversity), although the importance of diversity is less clear now.

The 5 species of animals represent differences in feeding types. Amphipods were intended as detritivors and grazers on filamentous algae that form a mat on the sand. Aberdeen omitted amphipods in some of their experiments, and did not seem to have very different results — although, of course, it was impossible to show effects on amphipods.

The decision to use non-aseptic sampling techniques, and to allow the microbial populations to be introduced passively with the animals, provides a lack of standardization of the non-algal microbial populations. Since we did not characterize these microbial populations, we do not know if they are responsible for any of the variability observed.

We have suggested that the microcosms could be used to study the survival, potential DNA transfer, degradation capabilities, and ecological effects of genetically engineered microorganisms.[37] The SAM has been used to study degradation of a toxic chemical; in spite of the disappearance of the parent chemical, the microcosms remained toxic to *Daphnia*, presumably because of a toxic intermediate.[38] Andrew C. Kindig[42] reports that a study of the ecological effects and DNA transfer of a unique microorganism is being conducted by Beak Consultants Ltd. for Environment Canada, SCC file No 066SS. Kel44-0-6238 (A Development of Methods to Assess Monitoring Techniques for Environmental Introduction of Microorganisms).

The ability to compute and statistically analyze data from such a variety of measurements was aided by the use of computers, originally a main frame, now microcomputers. By having the computer prepare summary tables of statistical differences (an upward arrow to indicate a statistically greater value and a downward arrow to indicate a statistically smaller value than the control), it was easy to view the treatment effects.[25] However, it has been more difficult to convey the information in a concise manner to an audience. In practice, we have tended to use the "BIG-THREE" (nitrate, algal volume, and *Daphnia*) or if time and space allow the "BIG-FIVE," (those three plus pH and oxygen gain), and to omit publishing additional data.

The extrapolation of results between microcosms and natural aquatic communities depends on the equivalence of the systems involved; these microcosms are designed to be like early spring communities, with high initial algal nutrients, and no significant food sources for the zooplankton apart from the results of primary production. No continuing algal nutrients are available without nutrient recycling. Communities with different community structures or limiting factors would be expected to behave differently. For example, small shaded stream communities are likely to be based on terrestrial plant and insects inputs, and the *in situ* recycling of algal nutrients would have little effect on food available to grazers. Similar caution would have to be exercised on extrapolating results to waters of different compositions, that might have very different pH or chelation potentials.

MISCELLANEOUS POTENTIAL PROBLEMS

Researchers have asked if the microcosms would give similar results if different species were substituted. In my experience, it is not possible to predict the outcome of microcosms with substituted species. In our laboratory experience, changing the genus of ostracods altered our ability to show increases due to streptomycin. In our original streptomycin studies, the ostracod *Cyprinotus inconguens* was used. *Cyprinotus* obtained only modest abundance in the controls, and established extremely dense populations in the streptomycin treated microcosms.[21] In subsequent experiments, we used the ostracod *Cypridopsis* sp. This genus established dense populations in the control microcosms, and as a result, we were not able to show differences as a result of streptomycin treatment.

The standardization of water was predicted to be a problem by several researchers, although it did not appear to have been a problem in these inter-laboratory tests. "Distilled water" has different meanings in different laboratories.[39] In ours, it was water collected and stored in Pyrex glass from a glass still with metal heating coils. Care was taken to allow adequate steam to escape during the distillation process to assure that compounds more volatile than water (e.g., ammonia) were dissipated. In many laboratories, distilled water is automatically passed through ion exchange cartridges, which may add significant amounts of dissolved organic materials, in part from microorganisms that may be living in the cartridges. These dissolved organic materials may contribute significant chelates and micronutrients. In other laboratories, distilled water is piped from a centralized still in metal pipes.

Overall, this work has demonstrated a degree of standardization of a multi-trophic, multispecies community. Additional work could increase standardization and the range of information provided. At their present level of development, the microcosms do provide information on the responses of lower trophic levels, and provide information on effects on food supply or water quality that may limit fish growth and survival. These microcosms offer the ability to confirm experimental results as subsequent times, or in other places. Their potential use can assist in the regulatory process as a means to determine if a test chemical is safe enough to use in the environment,[40] but will not provide all of the necessary information. The microcosms will provide information on competitive shifts within trophic levels, interactions between trophic levels, and the results of chemical transformations to more or less toxic chemical forms. The desirability of the test depends on the purpose at hand.

ACKNOWLEDGMENTS

The major development was supported by U.S. Food and Drug Administration Contracts 223-83-7000, 223-80-2352, and 223-76-8348. The U.S. Environmental Protection Agency (EPA), U.S. Army at Aberdeen Proving Ground, Maryland, and Marine Bioassays Laboratory, California are thanked for taking part in the inter-laboratory experiments. An EPA Peer Review Grant supported the modeling and timing study.[28] I thank the National Research Council and EPA, Athens Lab for supporting this manuscript effort and associated travel.

REFERENCES

1. Harrass, M. C. and P. G. Sayer. 1989. Use of microcosm data for regulatory decisions. In U. M. Cowgill and L. R. Williams, Eds., *Aquatic Toxicology and Hazard Assessment,* Vol. 12. ASTM STP 1027. American Society for Testing and Materials, PA, pp. 204–223.
2. Lampert, W. 1987. Predictability in lake ecosystems: the role of biotic interactions. *Ecol. Stud.* 61:333–346.
3. Sommer, U., Z. M. Gliwicz, W. Lampert and Anne Ducan. 1986. The PEG-model of seasonal succession of planktonic events in fresh waters. *Arch. Hydrobiol.* 106:433–471.
4. Taub, F. B. 1984. Synthetic microcosms as biological models of algal communities. In L. E. Shubert, Ed., *Algal as Ecological Indicators.* Academic Press, New York, NY, pp. 363–394, Chap. 13.
5. Conquest, L. L. and F. B. Taub. 1989. Repeatability and reproducibility of the Standardized Aquatic Microcosm: statistical properties. In U. Cowgill, Ed., *Aquatic Toxicology and Hazard Assessment,* Vol. 12. ASTM STP 1027. American Society for Testing and Materials, Philadelphia, pp. 159–177.
6. Conquest, L. L. 1983. Assessing the statistical effectiveness of ecological experiments: utility of the coefficient of variation. *Intern. J. Environ. Stud.* 20:209–221.
7. Crow, M. E. and F. B. Taub. 1979. Designing a microcosm bioassay to detect ecosystem level effects. *Internat. J. Environ. Stud.* 13:141–147.

8. Taub, F. B., M. E. Crow and H. J. Hartmann. 1980. Responses of aquatic microcosm to acute mortality. In J. P. Giesy, Jr., Ed., *Microcosms in Ecological Research.* Conf-781101; DOE Symposium Ser. 52, pp. 513–535.
9. Taub, F. B., P. L. Read, A. C. Kinding, M. C. Harrass, H. J. Hartmann, L. L. Conquest, F. J. Hardy and P. T. Munro. 1983. Demonstration of the ecological effects of streptomycin and malathion on synthetic aquatic microcosms. In W. E. Bishop, R. D. Cardwell, and B. B. Heidolph, Ed., *Aquatic Toxicology and Hazard Assessment: Sixth Symposium.* ASTM STP 802. American Society for Testing and Materials, Philadelphia, PA, pp. 5–25.
10. Taub, F. 1989. Standardized aquatic microcosm development and testing. In A. Boudou and F. Ribeyre, Ed., *Aquatic Ecotoxicology,* Vol. II. CRC Press, Inc., pp. 47–94, chap. 1.2.
11. Taub, F. B., A. C. Kindig, L. L. Conquest and J. P. Meador. 1989. Results of inter-laboratory testing of the standardized aquatic microcosm protocol. In G. W. Suter II and M. A. Lewis, Ed., *Aquatic Toxicology and Hazard Assessment,* Vol. 11. ASTM STP 1007, American Society for Testing Materials, Philadelphia, pp. 368–394.
12. Harrass, M. C. and F. B. Taub. 1985. Comparison of the laboratory microcosms and field responses to copper. In T. P. Boyle, Ed., *Validation and Predictability of Laboratory Methods for the Assessment of Fate and Effects of Environmental Contaminants in Aquatic Ecosystems.* ASTM STP 865. American Society for Testing Materials, Philadelphia, PA, pp. 57–74.
13. Rose, K. A. G. L. Swartzman, A. C. Kindig and F. B. Taub. 1988. Stepwise iterative calibration of a multispecies phytoplankton-zooplankton simulation model using laboratory data. *Ecol. Modell.* 42:1–32.
14. Swartzman, G., K. Rose, A. Kindig and F. B. Taub. 1989. Modelling the direct and indirect effects of streptomycin in aquatic microcosms. *Aquat. Toxicol.* 14:109–130.
15. Swartzman, G. L., F. B. Taub, J. Meador, C. Huang and A. C. Kindig. 1990. Modeling the effect of algal biomass on multispecies aquatic microcosms response to copper toxicity. *Aquat. Toxicol.* 17:93–118.
16. Swartzman, G. L. and K. A. Rose. 1983/1984. Simulating the biological effects to toxicants in aquatic microcosm systems. *Ecol. Modell.* 22:123–134.
17. Taub, F. B., A. C. Kindig and L. L. Conquest. 1988. Inter-laboratory testing of a standardized aquatic microcosm. In W. J. Adams, G. A. Chapman and W. G. Landis, Eds., *Aquatic Toxicology and Hazard Assessment,* Vol. 10. ASTM STP 971. American Society for Testing Materials, Philadelphia, pp. 384–405.

18. Taub, F. B., A. C. Kindig and L. L. Conquest. 1986. Preliminary results of inter-laboratory testing of a standardized aquatic microcosm. In J. Cairns, Jr., Ed., *Community Toxicity Testing*. ASTM STP 920. American Society for Testing Materials, Philadelphia, PA, pp. 93–120.
19. Taub, F. B. 1985. Toward inter-laboratory (round-robin) testing of a standardized aquatic microcosm. In J. Cairns, Jr., Ed., *Multispecies Toxicity Testing,* Pergamon Press, U.K., pp. 165–186, chap. 13.
20. Harrass, M. C. and A. C. Kindig. 1985. Responses of blue-green and green algae to streptomycin in unialgal and paired culture. *Aquat. Toxicol.* 6:1–11.
21. Kindig, A. C., L. L. Conquest and F. B. Taub. 1983. Differential sensitivity of new versus mature synthetic microcosms to streptomycin sulfate treatment. In W. E. bishop, R. D. Cardwell and B. B. Heidolph, Eds., *Aquatic Toxicology and Hazard Assessment: Sixth Symposium,* ASTM STP 802. American Society for Testing and Materials, Philadelphia, PA, pp. 193–203.
22. Taub, F. B., M. C. Harrass, H. J. Hartmann, A. C. Kindig and P. L. Read. 1981. Effects of initial algal density on community development in aquatic microcosms. *Verh. Internat. Verein. Limnol.* 21:197–204.
23. Taub, F. B. and M. E. Crow. 1978. Loss of a critical species in a model (laboratory) ecosystem. *Verh. Internat. Verein. Limnol.* 20:1270–1276.
24. Taub, F. B. 1989. Standardized aquatic microcosms. *Environ. Sci. Technol.* 23:1064–1066.
25. Taub, F. B. 1984. Measurements of pollution in synthesized aquatic microcosms. In H. H. White, Ed., *Concepts in Marine Pollution Measurements.* University of Maryland, College Park, MD, pp. 159–193.
26. Taub, F. B. and M. E. Crow. 1980. Synthetizing aquatic microcosms. In J. P. Giesy, Jr., Ed., *Microcosms in Ecological Research.* CONF-781101; DOE Symposium Ser. 52, pp. 69–104.
27. Harrass, M. C. and F. B. Taub. 1985. Effects of small fish predation on microcosm community bioassays. In R. D. Cardwell, R. Purdy, and R. C. Bahner, Eds., *Aquatic Toxicology and Hazard Assessment: Seventh Symposium.* ASTM STP 854. American Society for Testing and Materials, Philadelphia, PA, pp. 117–133.
28. Taub, F. B., A. C. Kindig, J. P. Meador and G. L. Swartzman. In press. Effects of seasonal succession and grazing on copper toxicity in aquatic microcosms. *Verh. Internat. Verein. Limnol.*
29. Anon. 1991. Standard Practice for Standardized Aquatic Microcosms: Fresh Water. *1991 Annual Book of ASTM Standards,* Vol. 11.04. ASTM E1366–90. American Society for Testing and Materials, Philadelphia, PA, pp. 1017–1051.

30. Taub, F. B. and A. C. Kindig. 1988. Standardized Aquatic Microcosm Protocol. SETAC Short Course, Modification of FDA Contract 223-83-7000 Final Report, Vol. II (1986).
31. Lintott, D. R. and S. Swanson. 1990. Evaluation of fate and toxicity of diclofop-methyl using an aquatic model ecosystem. Abstract O21, SETAC'90 Abstracts, p. 5.
32. Landis, W. G., N. A. Chester, M. V. Haley, D. W. Johnson, W. T. Muse, Jr. and R. M. Tauber. 1989. Utility of the standardized aquatic microcosms as a standard method for ecotoxicological evaluations. In G. W. Sutter II and M. A. Lewis, Eds., *Aquatic Toxicology and Environmental Fate,* Vol. 11. ASTM STP 1007. American Society for Testing and Materials, Philadelphia, PA, pp. 353–367.
33. Keating, K. I. and B. C. Dagbusan. 1984. Effects of selenium deficiency on cuticle integrity in the Cladocera (Crustacea). *Proc. Natl. Acad. Sci. U.S.A. (Ecology)* 81:3433–3437.
34. Keating, K. I. 1985. A system of defined (*sensu stricto*) media for daphnid (Cladocera) culture. *Water Res.* 19:73–78.
35. Meador, J. P. 1991. The interaction of pH, dissolved organic carbon, and total copper in the determination of ionic copper and toxicity. *Aquat. Toxicol.* 19:13–32. (See also Meador, J. P. 1988. The Chemical and Biological Factors Associated with Ecosystem Recovery from Copper Stress as Exemplified by Microcosms. Ph.D. thesis, University of Washington, Seattle, WA).
36. Sheehan, P. J., D. R. Miller, C. G. Butler and P. Bourdeau. 1984. *Effects of Pollutants at the Ecosystem Level.* John Wiley & Sons, NY. p. 443.
37. Taub, F. B. 1992. Synthetic microcosms as test systems for survival and effects of genetically engineered microorganisms. In Levin, M., R. Seidler and M. Rogul, Eds., *Microbial Ecology: Principles, Methods and Applications.* pp. 643–661.
38. Landis, W. G., N. A. Chester, M. V. Haley, D. W. Johnson, W. T. Muse, Jr., R. M. Tauber. 1989. Utility of the Standardized Aquatic Microcosm as a Standard Method for Ecotoxicological Evaluation. In G. W. Sutter II and M. A. Lewis, Eds., *Aquatic Toxicology and Environmental Fate,* Vol. II. ASTM STP 1007. American Society for Testing and Materials, Philadelphia, PA, pp. 353–367.
39. Specifications for Reagent Water. *1991 Annual Book of ASTM Standards,* Vol. 11.01. ASTM D1193. American Society for Testing and Materials, Philadelphia, PA.
40. Environmental Protection Agency. 1987. Toxic Substances Control Act Test Guidelines. Federal Register 52:36334–36371. (September 28, 1987).
41. Swanson, S. personal communication.
42. Kindig, A. C. personal communication.

CHAPTER 11

Seasonality of Sediment Toxicity in a Contaminated Harbour

R. Dermott and M. Munawar

ABSTRACT

Traditionally, sediment toxicity testing has focused on limited sediment collections and ignored the possibility of seasonal fluctuations in the bioavailability of contaminants. The results from Hamilton Harbour, Ontario, indicate poor survival of test organisms during the summer compared to the spring and fall seasons. Our sediment toxicity data have serious implications on the results of those investigations which are based on a single observation and do not consider seasonality of toxicity.

0-87371-845-3/93/$0.00 + $0.50

KEY WORDS

Amphipod, oligochaete, bioassay, oxygenation.

INTRODUCTION

The recent flurry of interest in determining the toxicity of in-place sediments to aquatic biota has resulted in the development of numerous bioassay techniques, a revised interest in the biotic index of the native communities, the development of the triage concept, and the multitrophic bioassay strategy to evaluate the impact of contaminated sediments.[1–3]

The evaluation of sediment toxicity is a complex subject. For example, the methods of sediment collection, sediment storage, and choice of assay organism, as well as genetic variability or hormesis in the test organisms can lead to complications in comparing mortality rates in different studies.[4–5] Sediment heterogeneity, water chemistry, and exposure temperature can result in conflicting results in related studies. Normally, sediment assays are conducted either once or over a limited period of time, with no indication of seasonal fluctuations in sediment toxicity. Previous work revealed that the sediments which were highly toxic during the summer of 1987 were non-toxic during late autumn and winter.[6] Therefore, it was imperative to assess toxicity from sediments collected from the same locations but at different seasons, particularly beyond the summer period of stratification from June until September.

In this paper, we focus on the seasonality of toxicity in an attempt to develop standardized testing procedures. Sediments from Hamilton Harbour, Lake Ontario were used to determine if (1) the acute toxicity of the harbour sediments prevent their colonization by invertebrates, or (2) the invertebrates could survive in the sediments which have become oxidized during the winter months.

MATERIALS AND METHODS

Hamilton Harbour, located on Lake Ontario, contains sediments that are contaminated with a variety of metals and polyaromatic hydrocarbons (PAHs) as a result of extensive metal smelting. The center of the harbour is 25 m deep, and due to the high anthropologic inputs to the bay, the water below 10 m depth undergoes anoxia during the summer months from July until September. As a result, beyond 6 m depth, the bottom fauna of the harbour is severely restricted to a few species of tubificid oligochaetes.[7] However, during the winter turnover, fish recolonize the deeper portions of the harbour.[8]

The sample sites were located by ranging with ship's radar. Sediments were collected at various times during the year in order to have sediment samples from spring overturn (April 1989, 1990), mid-summer anoxia (July 1987, 1989, and August 1989), fall overturn (September 1987, October 1990), and early winter (November 1989). Sediments were collected using a Ponar grab at three sites: Station 1 (depth = 11 m), Station 2 (23 m), and Station 6 (6 m) in the harbour (Figure 1). Three samples were collected at each site. The water overlying the sediments was removed by siphon and discarded. The top 10 cm of sediment from each of the three samples was removed and mixed together in clean 5-l apiary pails. The labeled pails were sealed, placed into coolers, and returned to the laboratory where they were stored at

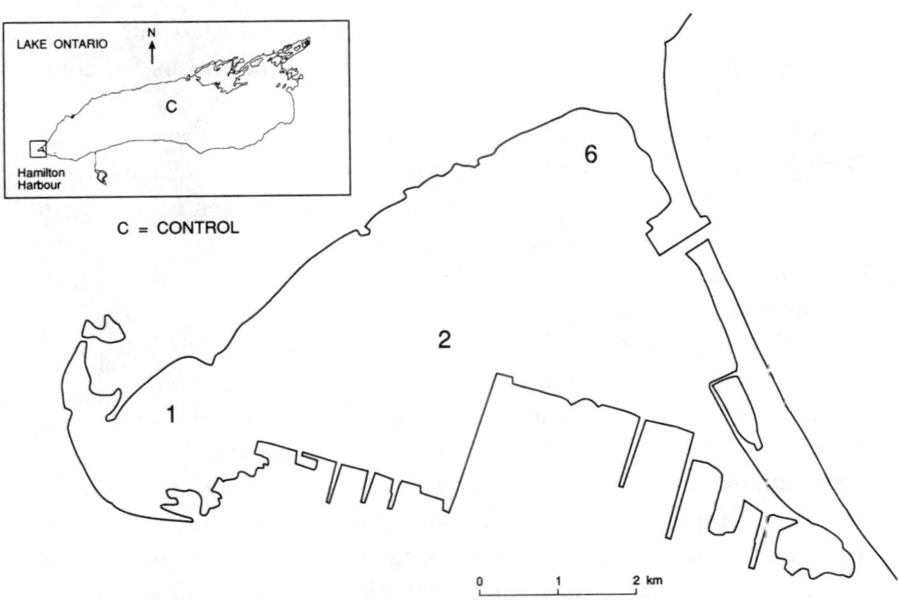

FIGURE 1. The site locations in Hamilton Harbour, Canada.

4°C until use within 2 weeks after collection. A control sediment was obtained from core samples collected near the middle of Lake Ontario at a depth of 125 m. The upper 15 cm of the cores were discarded, the lower sediment mixed and stored at 4°C until needed. This subsurface mud was deposited between 75 and 140 years ago. As the test organisms are detritivores, the organic content of the control sediment was supplemented by mixing into the mud 250 mg of tropical fish food per liter.

Toxicity testing was conducted in the various seasons with two benthic species, the pontoporeia amphipod (*Diporeia hoyi* Bousfield 1989), and the oligochaete (*Lumbriculus variegatus* Müller 1744). Both species can survive in sediments of lakes which have oxygen levels as low as 1 ppm. In this text, pontoporeia and lumbriculus are used as the common names of these two organisms. Pontoporeia were collected from eastern Lake Ontario at a site 35 m deep. The top 6 cm of sediment, containing most of the amphipods, was bagged with an equal portion of cold water, put into coolers, and returned to the laboratory. These amphipods were stored for 1 week at 10°C before use to allow the death of animals from sampling trauma. The lumbriculus were purchased from aquaria wholesalers and maintained at 20°C in 30-l aquaria on fine 100 μm silicon sand. The water was aerated with a slow water exchange of about 1 volume per week and the worms fed commercial trout food pellets.

For the assays, we used five replicate jars per test mud from each site. During each test, the subsurface core mud was used as a control sediment. To each 100-ml jar, 20 ml of mud was added resulting in a layer 2 cm deep. The mud was unscreened prior to use in order to limit the disturbance of the sediment particle size distribution and resulting contaminant mobilization. The mud was allowed to settle and surface oxidize overnight while under 1 mm of water. Later, the jars were gently filled with oxygen saturated lake water, so as not to disturb the sediment.

When required, the sediment containing the pontoporeia was passed through a 1-mm screen, the retained amphipods backwashed into a 1-l jar containing clean lake water at 10°C. Five immature pontoporeia (4 to 6 mm length) were transferred to each of the jars of sediment using a 2-mm diameter pipette. A mesh screening of 0.5 mm was placed over the mouths of the jars to prevent the organisms from escaping. The jars were randomly placed into 30-l aquaria, which were then filled with dechlorinated water to 4 cm above the jar mouths. A bubbler was then placed into each aquarium to oxygenate the water. After one week exposure at 10°C, the sediment in each jar was sieved through a 0.5-mm screen and the number of surviving amphipods was recorded. Details of the methods are published.[3]

The protocol for the worm assay involved passing a screen of 0.5-mm mesh through the disturbed sediment of the storage tank. The retained worms were backwashed into a white enamel dish. Ten medium to large (15 mm) worms were pipetted into each sediment jar and then placed carefully in a circle around the air bubblers in the aquaria. Water was slowly added to the aquaria until it just flowed into the jars, then maintained at 2 to 4 cm above the lips of the jars. Exposure of the worms was at 20°C. Oxygen readings were taken inside the jars after 48 h using an oxygen meter and again at the end of the incubation. After 2 weeks exposure, the sediment in each jar was screened and the worms present fixed in 4% formalin. The worms were then examined under a stereomicroscope in order to separate posterior worm fragments and native tubificid worms from the added lumbriculus. As *L. variegatus* reproduce by architomic division, only the anterior portions with developed prostomium were counted.

RESULTS AND DISCUSSION

All test sediments from Hamilton Harbour and Lake Ontario were fine silts with a dark color and low to moderate organic content (1.9 to 7%). Metal levels were greatest in the deeper harbour sediments, particularly those of Fe, Pb, and Zn (Table 1).

Throughout the experiments, oxygen levels in the test jars averaged 86% saturated with the lowest recorded oxygen reading being 5.3 ppm. Therefore the oxygen requirements of the test organisms were adequately maintained during the experiments. Survival in the control sediment averaged 96% of the introduced animals (Figure 2). There was no significant difference in the survival of the pontoporeia in the harbour sediments collected in April, as compared to the Lake Ontario sediment. On average, the highest survival in the harbour sediment occurred in the sediments from shallow site 6, where survival was 75% of that in the control sediment. No seasonality was evident in the survival of pontoporeia in sediment from this site or in the control sediment over the study period. Survival in the sediments from the mid-harbour sites 1 and 2, both from depths below 10 m, was strongly

Table 1
Contaminant levels in Hamilton Harbour sediments as acid extracted values μg g^{-1} except Fe and Mn. Metals were analyzed by the National Water Quality Laboratory, Burlington, Ontario, using atomic absorption spectroscopy

Location	Stn 1	Stn 2	Stn 6	Lake Ontario
Depth (m)	11.0	23.0	5.0	125.0
% organics	4.8	7.2	1.9	3.7
Cd	3.9	5.2	2.9	2.0
Cu	76.0	101.0	73.8	68.7
Fe[a]	14.2	23.2	10.7	10.4
Mn[a]	1.5	2.0	0.6	1.6
Ni	28.7	35.0	13.9	41.0
Pb	181.7	292.3	56.2	87.0
Zn	1425.4	1999.7	382.0	12.0

[a] mg g^{-1}

seasonal (G-test $p < 0.001$).[9] At both sites, poor survival occurred in sediment collected between June and August, as compared to the survival in the sediment collected from these sites during April, October, or November (Figure 2). Differences in the toxicity of the sediments were evident both in consecutive months at site 1, and between the same month in different years at site 2.

Similarly, survival by lumbriculus in the deep harbour sediments was also reduced during mid-summer (Figure 3). Seasonality in the toxicity of the harbour sediment was not as evident in the worms (G-test $p< 0.01$) as in the pontoporeia. In contrast, reproduction by the lumbriculus occurred over the exposure period, both in the control and shallow harbour sediment during mid-summer. As a result, the total number of lumbriculus in the jars after the exposure was normally greater than that added to the test jars.

Three factors prevent the establishment of a normal deepwater benthos in Hamilton Harbour. These are: (1) the elevated metal levels in the fine-grained sediments of the deeper portions of the harbour,[10] (2) the elevated toxic un-ionized ammonia levels in the water column,[11] and especially (3) the summer anoxia beyond a depth of 10 m.

The high levels of sediment contaminants, in particular Pb (< 200 μg/g), PAHs to 3.8 and the PCB levels above 0.1 μg/g would be expected to have adverse effects on the benthos of the harbour.[2, 8] In part, these factors are reflected in the limited benthic community over much of the harbour and very high levels of Fe and Mg found in the areas of the harbour.[7] When oxidized, the Fe^{+} ions protect the aquatic biota from exposure to toxic metals by the sorption of heavy metals into the

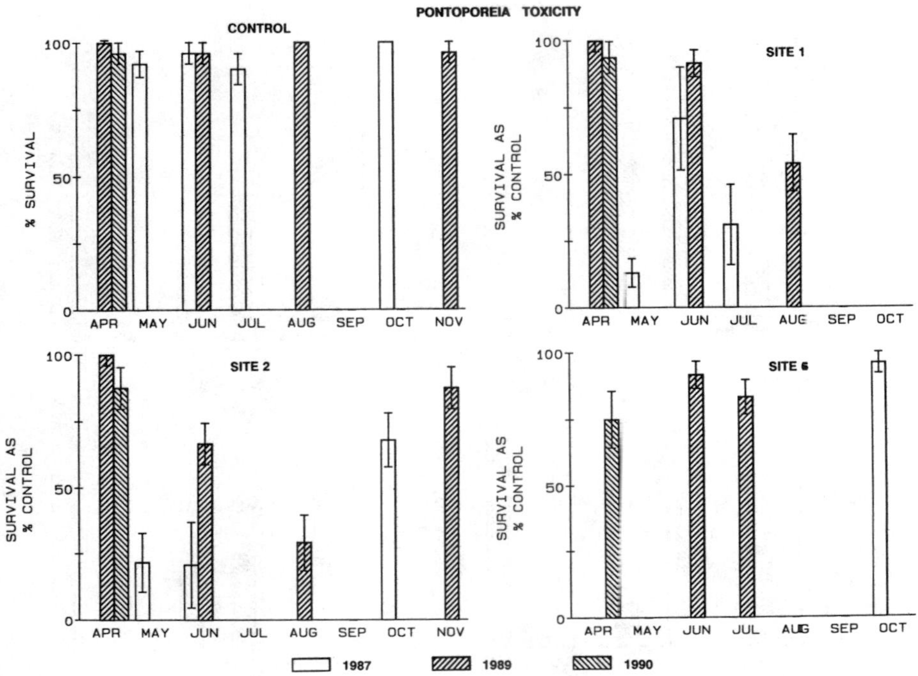

FIGURE 2. Seasonal changes in the percent survival of the amphipod *Diporeia* (Pontoporeia) in Hamilton Harbour sediment and survival in the Lake Ontario control sediment. Interval bars are ±1 standard error.

hydroxylated Fe.[12] This chelating action occurs during aerobic conditions at the sediment surface during the fall and winter overturn period (September to April), and would reduce the toxicity of the sediments to the benthos.

During anaerobic conditions of the hypolimnion in mid-summer, the chelating effect of Fe would be lost. As conditions at the sediment surface go anoxic, the redox potential (Eh), pH, and oxygenated valences of the metals in the sediments are changed. These all change the solubility, chemistry, and biological uptake of the metals, as well as the lower pH replacing the metal cations adsorbed on clay minerals with substituted H^+.[13] In addition, changing Eh and pH in the interstitial water influences the uptake of metals by the cell walls of the sediment microbes.[14] The mobilization and biologically more available metal cations in the summer collected sediments would increase their toxicity. This is only one of the causes of increased toxicity of summer sediments as compared to the spring or fall sediments (Figure 2).

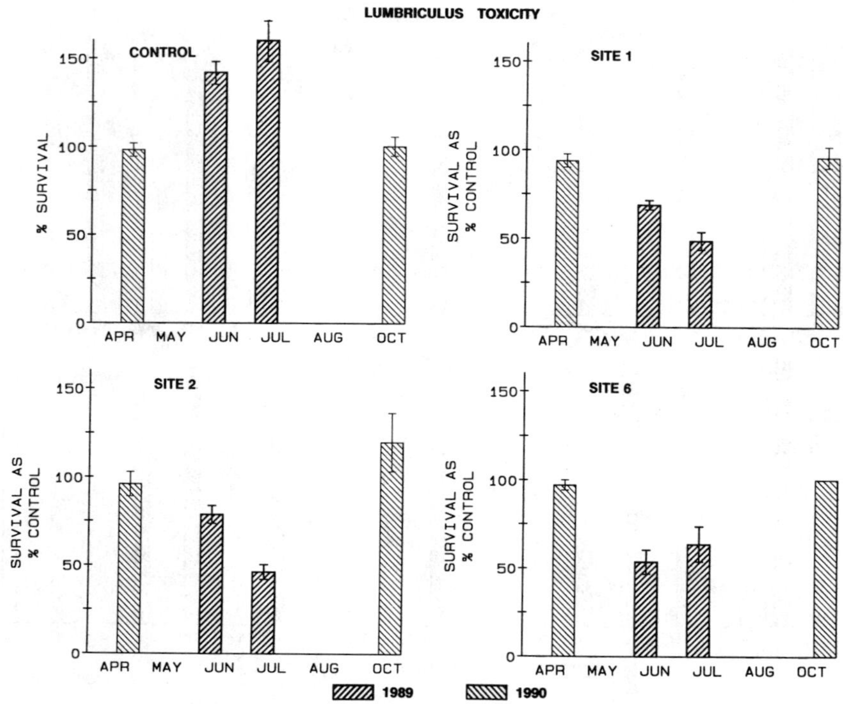

FIGURE 3. Seasonal changes in the percent survival of the oligochaete *Lumbriculus* in Hamilton Harbour sediment.

In addition to the influence of anoxia on metal solubility, the activities of the anaerobic microbes, particularly *Desulfovibrio* and *Clostridium*, in the summer sediments increase the levels of amines and H_2S in the interstitial pore water. At the redox boundary the amines are oxidized, liberating ammonia.[15] Toxic levels of un-ionized ammonia periodically occur in the water column of Hamilton Harbour. Levels often exceed 100 μg/L during the summer then drop below the 20 μg/L, IJC objective in the autumn.[11, 16] Briefly during May 1987, the un-ionized ammonia levels in Hamilton Harbour exceeded the LC_{50} levels of 300 μg/L.[17] During the same period, O_2 levels were below 7 ppm in the water column at sites 1 and 2, but remained above 10 ppm during May and June of the following year.[11] Survival of the pontoporeia added to the sediments collected during May 1987 was much lower than the survival in sediment collected in May and June the following year (Figure 2).

High levels of the ammonification and H_2S emitting from the anaerobic sediments collected during mid-summer would act synergist with the mobilized metal mixtures to increase the toxicity of the sediments. The assay procedure used attempted to oxidize the surface of the sediments added to the test jars by allowing the mud to compact overnight under a thin layer of water. A thin 1 to 5 mm surface layer of lighter oxygenated sediment formed prior to filling the jars with oxygen saturated lake water and adding the test organisms. The time of atmospheric exposure should be sufficient to allow volatization of toxic ammonia and H_2S from the sediment. The oxygenation period of the test sediments prior to adding the organisms may not be long enough to allow the hydroxylation of the Fe cations and chelation of the metals in the sediment pore water.

By the end of the experiments, the lower portions of the harbour sediments remained darker than the thin surface layer of oxidized sediment, mimicking the normal conditions occurring in organic rich sediments. This was the main reason why the sediments were not sieved prior to their use, as the screening would dilute the pore water during the settling phase, and alter the natural redox zones in the sediment.

In conclusion, the strong seasonal redox cycle of the sediments from Hamilton Harbour alters their toxicity to benthic organisms. Field collections and assays routinely conducted during the summer would conclude that invertebrates would not survive in the deeper portions of the harbour. However, the seasonal results from fall and early spring collections indicate that if sediments in Hamilton Harbour were to remain oxidized, benthic organisms could re-invade the hypolimnion. Therefore, these results have serious implications on the results of those investigations which are based on a single observation and do not consider seasonality of toxicity as demonstrated by the Hamilton Harbour sediments. It is imperative that the sediments be collected both during aerobic and anoxic conditions if they undergo seasonal fluctuations in oxygen levels. Only by assaying the sediments over at least three seasons can the results be compared using standardized sediment toxicity tests.

ACKNOWLEDGMENTS

We would to thank the following personnel who assisted in various aspects of the work: V. Hall, A. Aujla, and H.F. Nicholson.

REFERENCES

1. Nebeker, A.V., M.A. Cairns, J.H. Gakstatter, K.W. Malueg, G.S. Schuytema, and D.K. Krawczyk. 1984. Biological methods for determining toxicity of contaminated freshwater sediments to invertebrates. *Environ. Toxicol. Chem.* 3: 617–630.
2. Chapman, P.M. 1986. Sediment quality criteria from the sediment quality triad: an example. *Environ. Toxicol. Chem.* 5: 957–964.
3. Munawar, M., I.F. Munawar, C.I. Mayfield, and L.H. McCarthy. 1989. Probing ecosystem health: a multi-disciplinary and multitrophic assay strategy. In: M. Munawar, G. Dixon, C.I. Mayfield, T. Reynoldson and M.H. Sadar, Eds., Environmental Bioassay Techniques and Their Application. *Dev. Hydrobiol.* 54: 93–116.
4. Baird, D.J., I. Barker, M. Bradley, P. Calow, and A.M.V.M. Soares. 1989. The Daphnia Bioassay: a critique. *Hydrobiologia*, 188/189: 403–406.
5. Laughlin, R.B., J. Ng, and H.E. Guard. 1981. Hormesis: a response to low level environmental concentrations of petroleum hydrocarbons. *Science*, 211: 705–706.
6. Munawar, M. 1989. Ecosystem health evaluation of Ashbridges Bay environment using a battery of tests (M.I.S.A.). Unpublished report. Great Lakes Laboratory for Fisheries and Aquatic Sciences, Burlington, Ontario.
7. Johnson, M.G. and D.H. Matheson. 1968. Macroinvertebrate communities of the sediments of Hamilton Bay and adjacent Lake Ontario. *Limnol. Oceanogr.* 13: 99–111.
8. Harlow, H.E. and P.V. Hodson. 1988. Chemical contamination of Hamilton Harbour: a review. *Can. Tech. Rep. Fish. Aquat. Sci.* 1603. 91 pp.
9. Sokal, R.R. and F.J. Rohlf. 1969. *Biometry.* W.H. Freeman Company, San Francisco, 776 pp.
10. Poulton, D.J. 1987. Trace contaminant status of Hamilton Harbour. *J. Great Lakes Res.* 13: 193–201.
11. Barica, J. 1990. Ammonia and nitrite contamination of Hamilton Harbour, Lake Ontario. *Water Pollut. Res. J. Can.* 25: 359–386.
12. Lum, K.R. and K.L. Gammon. 1985. Geochemical availability of some trace and major elements in surficial sediments of the Detroit River and Western Lake Erie. *J. Great. Lakes Res.* 11: 328–338.
13. Strumm, W. and J.J. Morgan. 1971. *Aquatic Chemical Equilibria in Natural Waters.* Wiley-Interscience, New York, 780 pp.
14. Premuzic, E.T., M. Lin, H.L. Zhu, and A.M. Gremme. 1991. Selectivity in metal uptake by stationary phase microbial populations. *Arch. Environ. Contam. Toxicol.* 20: 234–240.

15. Stanier, R.Y., M. Doudoroff, and E.A. Adelberg. 1970. *The Microbial World.* Prentice-Hall Inc. New Jersey, 873 pp.
16. I.J.C. (International Joint Commission). 1978. U.S.-Canada Great Lakes Water Quality Agreement. Washington-Ottawa.
17. U.S.E.P.A. (Environmental Protection Agency). 1985. Ambient water quality criteria for ammonia — 1984. Rep. PB 85-227114.

INDEX